HIGH PROBABILITY

HIGH PROBABILITY

FINDINGS FROM A STUDY OF CE4, UFO, AND USO EXPERIENCES

LARRY G. HANSHAW, PH.D.

Table of Contents

Tables and Figures List

Overview

High Probability: Findings From A Study of CE4, UFO, and USO Experiences statistically analyzes data from close encounters of the fourth kind (i.e., CE4 events), and establishes how this phenomenon is related to other phenomena called UFOs (i.e., unidentified flying objects) and USOs (i.e., unidentified submersible objects). At the heart of High Probability is an intense focus on what abductees have said about their experiences in 216 cases which have occurred in over 30 countries. Basic and advanced statistical methods were applied to frequencies of selected variables to produce findings that reveal important relationships about CE4, UFO, and USO events. To enrich understandings about findings in this study, mathematical expressions were accompanied by proxy expressions such as unlikely and highly likely. Such expressions aided not only the understanding of symbols expressing the same meaning, but also provided readers with a way to appreciate the bigger picture of this research. This bigger picture throughout High Probability grew out of efforts to show the unique patterns embedded in the 216 abduction cases analyzed for this research. These patterns emerged in varying degrees in the form of frequencies of specific acts described by abductees. Additional frequencies from the accounts of observers who witnessed UFO and USO events provided frequencies of events that contributed to forming new understandings about these phenomena. Overall, this enabled the articulation of new correlations about the experiences of abductees and phenomena such as UFO and USO events. High Probability (HP) more than adequately addresses the claims of many sceptics whose views harshly criticize CE4 experiences and cast doubt on any experiences which might mildly suggest that an alien (or even avatar) is at the controls of unidentified aerial objects or that such objects are not of human design and control. With respect to CE4 experiences, specifically, Mack (1994) offered the following analysis of the fist-hand experiences of abductees:

> Efforts to establish a pattern of psychopathology other than disturbances associated with a traumatic event have been unsuccessful. Psychological testing of abductees has not revealed evidence of mental or emotional disturbance that could account for their reported experiences (Bloecher, Clamar, and Hopkins 1985; Parnell 1986; Parnell and Sprinkle 1990; Rodeghier, Goodpaster, and Blatterbauer 1991; Slater 1985; Spanos et al, 1983; Stone-Carmen, in press). (p. 16)

Yet, trained scientists ignore this type of evidence or they don't accept it but expect others with similar training and expertise to respect their decision-making behavior. The same can be said of skeptics, generally, including, of course, those who are also scientists. I believe it is far better to seek answers to challenging issues than to ignore a legitimate phenomenon that is akin to a public health issue which is sorely in need of scientific inquiry. Readers who are unfamiliar with the views of scientists and other experts in various fields who favor an open investigation into abductions and a resolution of who or what is flying around virtually unchallenged in the skies of this planet will learn much from reading High Probability. They may also want to know why there are so few scientific investigations reported in mainstream scientific literature. The absence of abundant physical evidence, as shown in High Probability, is not the correct answer. The paucity of research, however, involving what evidence is available deepens the complexity of the abduction phenomenon and continues to make an open discourse about abduction research in refereed journals and research-based books as ubiquitous as the last summit of alien beings at the United Nations. Readers of High Probability will be confronted with facts that point out that it was the theory-dependent reasoning of the Condon Report that convinced mainstream science that UFOs (and by extension CE4 and USO events) were not legitimate areas of scientific research (Hanshaw, 2004; Jacobs, 1992). Moreover, to not recognize world-wide abductions as a true case of the unfamiliar is no less than poor vision in need of surgery or new optics or both (Cherniack, 2010). Put simply, part of the goal of this research is to answer a new question: *What if abductees have been telling us the truth all along?* I suppose my background in chemistry and mathematics and many years of guiding graduate students through all sorts of dissertation topics has made me qualified and curious, still, to look into the abyss and ask: Who (or what) goes there?

There is a dominance of probabilities presented throughout this book which all but eliminate chance as the best explanation for various hypotheses. These probabilities paint a very provocative picture of the abduction phenomenon and its relationship to UFO and USO experiences. A reset of this magnitude could change mind sets across many communities of thought regarding abductions as well as UFO and USO events. High Probability is not intended to entertain any audience by making light of what many scholars and scientists consider to be a serious matter. Instead, HP raises the level of interdisciplinary discourse about the three aforementioned phenomena by providing a wealth of evidence that (1) agrees with previous findings of other researchers and (2) goes beyond what is widely known with new relationships supported by findings that clearly show that chance was not the favored explanation for a number of the hypotheses investigated. This point cannot be overstated. The examples of research results presented below are part of a larger body of such results that are either not discussed at all in other texts or when they are discussed, such results are not presented at the same level of specificity as offered in High Probability (i.e., see, for example, Comparison to Other Works, p. 15). In consideration of another perspective, the overall content and conclusions of High Probability will serve as a shoulder of support for the unknown numbers of experiencers some of whose accounts have made this research, and similar research on abductions, possible in the first place. Your cooperation is appreciated, especially because of the uninformed criticisms which seem to be attracted to such brave decisions. A very careful examination of available data about alleged human abductions by alien beings is a reservoir of information that can help researchers arrive at the truth about controversial phenomena. In fact, when abundant physical evidence is hard to come by, a researcher should be encouraged by what forensic specialists in law enforcement tell us—though often microscopic, at every crime scene there is always some type of evidence left behind. With regard to the abduction phenomenon, the trail of physical evidence has included such things as DNA evidence of body hair from an intimate encounter between a human being and an alien abductor; and (2) the bodies of living abductees and what they say are changes to their bodies for which there is no previous history or logical explanation (i.e., cuts, punctures, implants, or in quite a few cases, missing fetuses that defy previous results from a physician's earlier medical examinations(s) where everything was normal, but after ruling out the usual suspects that might cause the loss of a child, a fetus ends up missing). Hard, perhaps, to wrap one's mind around, but this is just the tip of the iceberg regarding what some abductees say they have experienced. Believe it or not, these types of outcomes are found in the literature, world-wide, and vetted in numerous cases in reports of trained investigators, physicians, law enforcement authorities, psychiatrists, psychologists, and other professionals seeking an understanding of what abductees have reported. In my opinion, the abductee in a close encounter of the fourth kind is both a witness and a victim to a crime (by any name or circumstance) and should be valued as a victim with information of help to investigations and not dismissed or called crazy or a liar for the story they desire to understand and to share in hopes of helping themselves and others. After all, there is the possibility that these people are telling the truth! Consider this: It is extremely unlikely that everyone, without exception, who claims to have been abducted is crazy or lying or, well, both! Critics, especially, should realize that saying such things about everyone whose experience they doubt is not proof! On the other hand, High Probability lives up to its end of the debate by providing research that explores the correlation between (1) states of mind (i.e., awake, dream/sleep, and hypnotized; see Chapter 5 of High Probability); (2) abduction-revealed acts as a consequence of abduction events; and (3) two other selected variables key to describing aspects of abduction events. Items 2 and 3 are reported in Chapter 3. There is also relevant research presented in Comparisons to Other Works in the Field which dismantles the regularly cited claims of critics that abductees are fantasy prone and tend toward false memories.

Overall, this overview of High Probability only highlights some of what was found in a detailed analysis of accounts involving CE4 experiences and their relationship to UFO and USO events. For example, additional information characterizing the uniqueness of High Probability is presented below in Items 2 and 3. Other general information about key aspects of this study are outlined more fully in the Table of Contents: Chapters 1-8. Facts about the author are presented in the curriculum vita at the end of this book (…just in case you are curious about what I did between 1965 and 2019).

The uniqueness of this book is that it examines relationships between selected variables and uses frequencies of these variables to determine if a given frequency is greater than the frequency expected by chance. This powerful approach helps to define what is and what is not considered to be the best explanation for a given hypothesis explored. Since the data analyzed in high probability was extracted from investigated cases involving abductees, there is little between the witnesses to these events and the perpetrators of these events. This is why the state of mind of abductees is an important area for expert opinion to weigh-in on to appropriately decide important issues on the basis of research-based findings and not on theory-dependent reasoning. The consistency of the narrative told in this research is in large part due to the kinds of data-driven findings in the five examples below:

(1) It is highly likely that UFOs may be involved in human abductions, since the frequency of UFO sightings in the same range of locations as CE4 cases is highly significant across the four ranges of G. P. S. Values examined: Pearson Chi Square = 10.759, p = .013. The effect size for these findings (Phi/Cramer's V = .163) was small, but significant (p = .013) for both measures. Thus, both measures indicated that to a small, but significant degree, the null hypothesis was false.

(2) The shapes of craft associated with CE4 cases and USO cases were not found to be significantly related, since the frequencies of sightings of USO craft shapes and reports of abductions involving CE4 craft shapes in the same G. P. S. Location ranges were not significant: Pearson Chi Square = 6.555, p = .088. Both effect sizes were small (.159) for Phi/Cramer's V and both measures were insignificant (p = .088). These results imply, however, that in an abduction scenario, it is just as likely that a CE4 craft shape will

be involved as might be a craft shape associated with USO events. This means that the results simply did not favor the frequency of either CE4 or USO craft shapes as the case types contributing most to the outcome of a test of the relationship between their frequencies.

(3) In an examination of the relationship between UFO craft shapes and USO craft shapes and the extent to which they share common locations over four G.P. S Location Ranges, the results indicated that location ranges common to craft shapes of both case types were found to be highly related: Pearson Chi Square = 19.137, p = .000 (p < .0005). Thus, in areas where CE4 events have occurred, the relationship of craft shapes of UFOs and USOs across the same G. P. S. location ranges indicated that UFO and USO craft shapes may also be involved in close encounters of the fourth kind, given the highly significant relationship and effect sizes they share with the craft shapes and locations of cases involving CE4 events. Both case types have craft shapes that are indistinguishable from the shapes abductees describe in CE4 cases. It may also be that following an abduction, and leaving the scene, either of these case types might be identified but not necessarily associated with an abduction. This important observation might apply more to UFOs than USOs, since more sightings of UFOs occur over land than craft identified as USOs which are identified in events involving seas, oceans, lakes or other water ways.

(4) High Probability identified the craft shapes that appear most across four G. P. S. ranges of location values world-wide. The (2.0-4.0) range of G. P. S. Values was the corridor with the highest frequency of craft shapes from all case types most contributing to the significant results: Pearson Chi Square = 20.856, p = .002. The effect size of the relationship across location ranges was small for Cramer's V and moderate (.211) for Phi. The effect size was significant for both effect size measures (p = .002).

(5) Using descriptions from 172 of 216 men and women, it was found that the frequency of descriptions across six alien types was greater than the frequency expected by chance: Pearson Chi Square = 475.472, p = .000

(p< .0005). This result favored the Humanoid species (i.e., greys as the type most contributing to the overall outcome). It is also worth noting that across the 216 people in this study, many of these persons saw at least one alien type while other abductees either interacted with or simply saw more than one species during a single abduction event. Other interesting follow-up findings (from Chapter 1) indicate that aliens tended to work more in the alone condition (i.e., Pearson Chi Square = 9.412, p = .002) than in the together or cooperative condition when abducting humans and that the alien types significantly preferred telepathy as their chosen style of communication (Pearson Chi Square = 7.000, p = .008). Readers should note that whenever the number following (p =) is equal to or less than (.05), the result is significant. The lower such values get, makes the interpretation of results lean more toward being a highly significant outcome.

Comparisons to Other Works in the Field

Recently published books such as Paranormal Claims: A Critical Analysis (edited by Bryan Farha) and Aliens, Ghosts, and Cults: Legends We Live (Bill Ellis) are examples of views among skeptics that either directly or indirectly conflate all cases involving UFOs, CE4s, and USOs with fire-side tales or pseudoscience or both. Such broad strokes fail to recognize the variety of evidence as found in High Probability (HP) as well as older sources that that are both technological (i.e., radar-visual, photographs, trace evidence, drawings) and biological in nature (i.e., scars, implants, missing fetuses, scoop marks). Researchers who make themselves aware of what has happened with regularity in abduction cases soon realize that not all types of the evidence just mentioned (there are other types) occur in each case. Nevertheless, the analysis of a large enough body of cases soon reveals the patterns that are evident in the findings of High Probability. What separates High Probability from other books about CE4 experiences, particularly, is High Probability's incorporation of research findings from, for example, psychology (Lynn & Kirsch, 1996) and psychiatry (Mack, 1994). These experts, among others, respond to claims of critics and others who contend that abductees are fantasy prone and tend toward false memories. Analyses of Susan Clancy's recent book Abducted: How People Come to Believe They Were Kidnaped by Aliens (Harvard University

Press, 2005) is an example of how theory-dependent reasoning masquerades as a supposedly serious discourse about abductees and the abduction phenomena (Jacobs, 2006; Hopkins, 2005). In contrast, High Probability utilizes investigator vetted sources from which data was extracted and analyzed to establish multiple patterns of consistency that agree with previous findings (i.e., Bullard, 1998; Jacobs, 2003; Hopkins, 1981). High Probability presents a mathematical approach to framing specific answers to old questions which before now have relied upon a mostly prosaic approach to explaining underlying relationships among CE4, UFO, and USO phenomena.

Offering a fresh analytical approach, HP provides unique insights into key issues common to CE4, UFO, and USO phenomena that also will address the general public's desire to know if we are alone in the universe. The findings and conclusions throughout HP also support the search for truth across the research agendas of abduction researchers and organizations across UFO and USO research communities. In response to whether or not we are alone in the universe or to address some of the issues related to who might be in the galactic neighborhood, HP uses mathematics to find relationships between variables that may lead to more specific findings of causality. A step even higher would be to amass, whenever possible, physical and biological evidence such as the clothing of abductees that might contain fluids, skin cells, or hair from contact with alien captors. An abundance of such evidence might support DNA analyses that could create an entirely new discourse about CE4 experiences and their relationship to UFOs, and USOs. Presently, however, HP presents *unlikely* and *highly likely* findings about interrelationships embedded in CE4, UFO, and USO experiences that offer a very different perspective of a world-wide phenomenon.

HIGH PROBABILITY

FINDINGS FROM A STUDY OF CE4, UFO, AND USO EXPERIENCES

LARRY G. HANSHAW, PH.D.

Author Note

Larry G. Hanshaw is Professor Emeritus of Secondary Science Education. Correspondence concerning this book should be addressed to Larry G. Hanshaw, Ph.D., P. O. Box 1568, Oxford, MS 38655. Email: lhanshaw@bellsouth.net

CHAPTER 1

An Analysis of Experiences Defined as Close Encounters of the Fourth Kind

C. D. B. Bryan defined *Close Encounters of the Fourth Kind* as a category in which personal contact between an individual or individuals is initiated by the "occupants" of the spacecraft. Such contact may involve the transportation of the individual from his or her terrestrial surroundings into the spacecraft, where the individual is communicated with and/or subjected to an examination before being returned. Such a close encounter is usually of a one-to-two-hour duration. (1995, p. 9)

Over the years, UFO encounters and alien abduction accounts have presented a mystery in search of an explanation (Bullard, 2010). I was fascinated, for example, by the 1961 experience of Betty and Barney Hill (Fuller, 1966) and the television program, *The UFO Incident.* The program aired on NBC, October 20, 1975 (Clark, 1998) and both versions of the story revealed the fear, disbelief, and the need for professional assistance now so often associated with abduction events (Fuller, 1966; Hopkins, 1981; Mack, 1994). Nevertheless, I wondered if such things really happened and how events of this nature might be assessed along some relativistic scale such "as "more likely" and "less likely," or "in terms of probabilities" (Sturrock, 1999, p. 164). Whatever is occurring will require that further research be performed to establish the truth about alien abductions

(Donderi, 2013). Moreover, available evidence (i.e., human observations, physical trace evidence, abductee accounts, photographs of craft, radar records, abductee-drawn illustrations, and regression hypnosis) must be examined for veracity and patterns in relationships to establish a reasonable foundation for arriving at the best explanation of this world-wide phenomenon. Although the field of ufology has contributed to keeping abduction stories as near to the mainstream of scientific thought and public awareness as possible, the battle has not been largely successful. Many mainstream journals and news outlets still balk at the idea that there is a genuine problem of scientific interest imbedded in the study of UFOs, USOs, and abductions, including the U. S. government and various independent sources of research funding. Nevertheless, science and mathematics principles offer important ways to "recognize some substance beneath the clutter" (Bullard, 2010, p. 24; Hynek, 1972). For example, one approach employed to express our hope that we are not alone involved sending probes into space bearing human images and musical selections put onto a gold anodized disc, complete with instructions for how to play back the recording (Jet Propulsion Laboratory, 2003). While we have not yet heard a response via this approach, an alternative approach offering clarity to the above question suggests the possibility that we already may have received a response many times over. Perhaps, the world-wide sightings of spacecraft and the occurrence, in some cases, of abduction events are parts of the answer we have been seeking all along. Perhaps, those we thought to be intelligent enough to play back our music and understand our genetic code, indeed, are intelligent; except, they have come to see for themselves instead of replying via radio signals or by some other technological means (i.e., sending back their own genetic code, or music, or location in the cosmos). Maybe their approach has been to obtain answers about us which they desire and/or to communicate with humans in ways that are not in keeping with our logic and/or expectations, but more in keeping with their own. Others, researching UFOs and, in particular, abductions, seem to agree that extraterrestrials may have taken a more direct (and personal) approach. "It is possible that aliens have been abducting humans for hundreds of years, but we have no evidence for it. We have collected some reports suggestive of abductions in the 1900s, and we have investigated abductions that occurred in the early 1930s. The main bulk of the abductions seems to have begun in the mid-twentieth century, perhaps coinciding with the first UFO sighting waves in the 1940s. Our knowledge is limited by the age of the people who explore their experiences; most abductees who have come forth in the last five years are under sixty years old" (Jacobs, 1992, p. 309). Still others suggest patience is in order: "Do not prejudge. Realize that if any aspect of the UFO phenomenon as reported is true, then any of the rest of the reported phenomena may be true too. Try not to put anthropomorphic limits on what may be an entirely alien intelligence and technology. The true skeptic cannot, at the beginning, accept the impossibility of anything" (Hopkins, 1987, p. xiii). Lastly, there are clinical findings (Laibow, 2016) that indicate significant discrepancies between observed and expected data involving "patients who believe themselves to be UFO abductees..." (Laibow, 2016, p. 1 of 10). In summary, the author found (1) "an absence of major psychopathology" among abductees; (2) that "certain details of the scenarios repeat themselves with disturbing regularity no matter what the educational, national, social, experiential or other demographic characteristics of the reporter"; (3) "If their stories (i.e., from abductees who were awake and not initially hypnotized; parentheses my own) were substantially different from the concordant abduction scenarios produced under hypnosis, a different phenomenon would be taking place"; and (4) that while abductees show a kind of trauma akin to PTSD (i.e., post-traumatic stress syndrome), the similarity is not unquestioned. That is, "If the abduction scenarios represent only a fantasy state, then it is worth investigating why (and how) this particular highly concordant and deeply disturbing fantasy is involved in the pathogenesis of a condition otherwise seen only following externally induced trauma" (Laibow, 2016, pp. 2-8 of 10; Appelle, 1995/96). It appears then that derisive terms used to describe abductees is little more than juvenile name calling and a dismissal of experiences of abductees is not based upon findings from sound investigations. Using information from abduction cases analyzed and presented in this research, further evidence may be established that, in turn, increases the scientific understanding about abductions; a phenomenon, the truth of which, we currently only partially know or understand.

Purpose

The primary purpose of this research was (1) to explore relationships between data sets representing descriptive elements of close encounters of the fourth kind as described by individuals who say they were abducted by non-human biological entities or extraterrestrial beings; (2) to investigate four research questions and hypotheses concerning the Abduction Phenomenon (Bullard, 1998) and additional issues related to it; and (3) to present several new perspectives arising from an analysis of data on craft shapes and G. P. S. data which together produced new insights defining the relationships found between CE4 experiences and UFO and USO events. A secondary purpose of this research was to present a more detailed picture of the apparent relationship between alien beings who work alone and cooperatively during abduction episodes. The use of total abduction scores quantified the experiences of abductees and made this aspect of High Probability especially informative. These scores were derived from 216 abduction cases that met case selection criteria developed during research for this book. Moreover, the calculation of percentile rankings made possible the emergence of insights that point out which alien beings worked alone, and which ones worked cooperatively with other beings during abduction events. During these events, humans were the only witnesses present. It is therefore important to consider the value of what can be learned from efforts by civilized countries all over the world when an earnest effort is made to collect, debate, and analyze abduction data so that a body of reliable information, over time, might help establish a framework for informed actions. Determining who (or what) is doing what and to whom applies to all sorts of human activities. Logically, such behavior should apply as well to UFO and USO sightings. Even though there is not a wealth of hard evidence, there is some hard evidence upon which to base sound investigations that might lead to more definitive answers. What is needed is more physical evidence to accompany what is already in hand: implants that don't get rejected by the body's immune system; craft photographs; DNA evidence (i.e., from alien hairs and ancient skulls), biological/chemical residue left on clothing following examinations by alien beings, radiological readings, abductee and non-abductee drawings, radar-visual records, and chemical trace element tests of soil. There is a need to do more. For example, should abductees be tested to detect any chemical imbalance present due to consequences of prolonged space flight? This is routinely done to astronauts. If more were done to add to what is available evidence, over time, more analyzed information, patterns, and evidence of many types could reach a critical mass that, in turn, might likely be beneficial to a greater understanding of CE4 events. Findings reported as *not likely* and *highly likely* may help to shape the public's understanding of the truth about the abduction phenomenon and likely related phenomena as well (i.e., UFOs and USOs). High Probability is, therefore, a conversation about controversial phenomena conducted in words and mathematics in order to construct a new precipice from which readers can view many relationships that connect CE4, UFO, and USO events in ways that all but eliminate *chance* as the preferred reason for such relationships. It is well understood that for some people only an alien entity that is interviewed on the evening news will be enough proof to end their wait to discover if intelligent life exists beyond the earth. For others, however, their wait ended the moment of their first abduction. This research hopes to make their experiences more widely known and better understood in relationship to UFO and USO events. A non-prosaic analysis of these relationships is long overdue.

Resolving a Tripartite Issue

As implied in C. D. B. Bryan's definition, close encounters of the fourth kind do not occur in a vacuum. Craft of unknown origin and, presumably, beings connected in some way with such craft are, logically, connected. That is, the craft are likely the result of some manufacturing process, like automobiles here on earth, and, therefore, do not come from raw materials to finished product on their own; beings of an intelligent nature are also likely involved. Hence, an inextricable relationship likely exists between (1) non-human intelligent beings, (2) their craft, and (3) a combination of technology, material science development, and propulsion engineering that is beyond anything man has so far developed. Therefore, this chapter and Chapters 2, 4, and 7 will present findings about the inextricable relationship described above in order to achieve an even greater proximity to the elusive "complete explanation" (Appelle, 1995/96, p. 65) that will seamlessly integrate UFO, CE4, and USO experiences into an understandable whole. So far, craft of various types have been observed on land and over various bodies of water. Sightings widely reported by major networks and newspapers on May 28, 2019 probably come to mind. Many craft shapes similar to those in UFO and USO cases also have been described in abduction accounts. Persons known as contactees have observed lights, buttons, and consoles within ships they claim to have entered or were invited. With respect to abductees, however, much more happened. It is their experiences which are explored in High Probability in order to extract the kind of evidence that bears directly on the observed craft involved in abductions. These cases also will speak to the technology displayed by such craft and, most importantly, the various types of beings reported aboard such craft. Currently, the abduction discourse is somewhere between *dismissed out of hand* and *open for further study*. Moreover, the shared variance between the possibility of intelligent control of UFOs by alien beings and abduction accounts requires that additional evidence be obtained. "So far, the strongest evidence is the myriad of abduction reports that have surfaced, with the congruence of narrative and the richness of exact detail. "Hard" evidence has been slow in coming, but is increasing... In the long run, the hard evidence may be the most important supportive evidence, but currently it is the physiological and psychological effects of abductions that provide many of our strongest clues to the abduction mystery" (Jacobs, 1992, pp. 239-240). The 1988 Khoury Abduction (Chalker, 1999) which readers should examine is an example of the type of *hard* evidence referred to and more is needed.

Although it is beyond the scope of this book to define beyond *any doubt* exactly whose technology is responsible for the descriptions given by abductees, it is still within the scope of this research to present findings that fit somewhere between *not likely and highly likely* in order to remove as much doubt as possible from the relationships shared by UFO, USO, and CE4 events. Many accounts in this study revealed (1) interior and exterior details of craft aboard which abductees were taken; (2) movement without walking; transportation through walls and closed windows all while encased in a wiggling tunnel of light; (3) musty smells; unusual lighting; furniture configurations that included tables and the number of legs; levels of humidity experienced during abduction episodes; and (4) low, humming, turbine-like sounds. However, not all abductees mentioned the same identical things. When was the last time you had a dream, woke up, and specifically remembered how much moisture was in the air where your dream took place? Such things would be easier to recall, if you were not dreaming in the first place. Perhaps, an increase in the analysis of CE4 cases will thwart any tendency to dismiss the unfamiliar and, thereby, simultaneously embrace a scientific desire to know the unknown. For example, new planets formerly thought not to exist, as well as the biology they might harbor, are subjects currently being investigated (Hanshaw, 2004). Given the age of our sun (13 billion years) and the time frame for development of intelligent life here on earth (some 4.5 billion years), it is conceivable that some other civilizations may be a billion or more years more advanced than life on this planet (Cherniack, 2010). Technology here has taken us to the moon and back, since the Wright brothers flew at Kitty Hawk in 1903. Given the billion or more years lead-time that some other possible civilizations may have over earth-based technology, it is not unreasonable to suggest that *they* may have solved problems of interstellar flight that we can only conceive of in terms of fantasy adventures like those seen in today's science fiction films. If abductees are to be believed, then serious thought must be given to the possibility that just because mankind has not achieved interstellar flight does not preclude the possibility that other possible civilizations have already achieved mankind's dreams. After all, it seems illogical that abductions would occur involving detailed descriptions of abductors and their craft and no one would care enough to pay serious attention to such accounts, or the persons involved. There is a paucity of persons willing to come forward to share their abduction experiences; criticism-free. Everyone who does come forward to share the unfamiliar *must* be delusional in some respect, right? How perfect is that?

The Abduction Phenomenon

The abduction phenomenon is defined by the diversity of experiences described by human beings who say they encountered alien beings and were taken aboard a UFO. Each of the **216** cases collected for this study were analyzed using the five tables of descriptive information by Bullard (1998). Table 2 in Bullard's "Sequence of episodes" describes "the abduction story...in eight possible episodes—capture, examination, conference, tour of the ship, journey or other-worldly journey, theophany, return, and aftermath" (Bullard, 1998, pp. 8-9). Tables 1, 3, 4, and 5 in the "Abduction Phenomenon" describe additional aspects of Bullard's research employed here to support the purposes of this research. For example, the 167 elements (i.e., headings, sub-headings, and components of each) within the five tables referenced above were adopted and combined with latitude and longitude data (not part of Bullard's original work), to produce a *screening mechanism* to collect abduction event details congruent with both *sequence of events* elements and any of the additional elements described in the four tables mentioned above. Information in the five tables also were used to achieve the purposes of this research as mentioned earlier. In summary, the sequence of events devised by Bullard have the following titles and descriptive elements within each table as shown by the number in parenthesis (i.e., except the latitude and longitude elements in Bullard's Table 1 below):

(1) Table 1: Location and Duration of Abduction... (plus added data)...(5) elements;

(2) Table 2: Sequence of episodes of the abduction event...(84) elements

(3) Table 3: Description of beings' appearance and behavior....(45) elements

(4) Table 4: Description of the craft....(19) elements

(5) Table 5: Description of mental and physical control...(14) elements (Bullard, 1998, pp.7-9; 12; and 14).

The five tables as described above were used to produce an *Abduction Phenomenon Total Score*. This score is a frequency representing the total of all table elements judged to be present in a given abduction account. Abductions involving more than a single person in a single event generated a separate report for each abductee, since persons may not experience identical episodes, even though the order of episodes remains the same for any abduction event (Bullard, 1998). Headings and subheadings across the five tables describing an abduction story were countable elements in this approach, since doing so added details and clarity to what abductees say they experienced. Thus, an Abduction Phenomenon Total Score could range from zero to 167. These scores were used to evaluate research questions and hypotheses presented later in this study.

It is conceded that no checklist, survey, or questionnaire ever constructed perfectly anticipates any subject matter. Therefore, concerns about the reliability (and validity) of an instrument are paramount, if the data collected is to be valued.

With respect to the "Sequence of episodes" contained within the Abduction Phenomenon, Bullard did not offer the eight episodes as a *data collection tool*. The adoption of the Sequence of episodes as a screening device was a decision by this researcher to use the tables of information Bullard described as a vehicle to collect information from abduction accounts read for this research. Because of Bullard's scholarly work, the five tables comprising the sequence of episodes made it possible to derive information that could be compared to pre-existing data. Additionally, an estimation of the reliability and validity of this *scholarly work turned checklist* was also reasoned to be high. This is because there was hardly any difficulty experienced at being able to find a reasonable match between words representing what abductees experienced (i.e., more so than what reporters thought/believed happened) and the wording of various elements in the eight episodes of the sequence of events. This congruence (i.e., content validity) produced the frequency count or score for each of the 216 cases reviewed in this study. Readers may wish to consider the results of the Sequence of Episodes Analysis (see Table 1) which convey reasonable support for the presence of an orderly consistency of events occurring during abductions. Moreover, results of the analysis also should allay concerns about any account details that may have been overlooked or omitted (i.e., a possible influence on scoring results). After reading hundreds of accounts leading to the 216 experiences in this research, it was concluded that the eight categories in the sequence of events were very effective at helping to produce trust worthy scores (i.e., high reliability), even though an understandable number of details germane to the overall abduction experience were found to be irrelevant to the eight components comprising the sequence of episodes.

Abduction Case Selection Criteria

The following criteria guided the selection of abduction cases included in this research: (1) Name or pseudonym, if given; (2) *Location (city, state, region of country)*; (3) *Age of experiencer*; (4) *Gender*; (5) Regression transcripts; richly detailed questions/answers presented by experienced UFO investigator/hypnotist; reporter; (6) Unbiased reporting of experiences/accounts; (7) *Events investigated/ reported by respected ufologists; ufological organization(s); law enforcement; other professionals*; (8) *CE4 event*. All elements in italics identify the minimum criteria required for cases in the literature to be included in this study, while all other elements were greatly desired, if possible. Hence, all cases included in this study met a minimum of five out of eight of the criteria listed above. Even though a few cases were single witness cases, they were determined not to be a hoax by investigating organizations (i.e., local police, ufological organizations, medical professionals, military officials, or other professionals).

Sequence of Episodes Analysis and Research Question 1

A Sequence of Episodes analysis was performed to determine if the same relative pattern, as described in Bullard's *Abduction Phenomenon* (Bullard, 1998, pp. 1-26), could be found in cases examined in this research. A question format (adapted from the above work) was used to perform the Sequence of Episodes Analysis. The results shown in Table 1 became the response to the following aspects of **Research Question 1.** What *percent* of the 216 cases:

(1) "featured all eight abduction episodes?; (2) indicated that *capture* was evident?; (3) indicated that *examination* followed capture?; (4) indicated that *conference* followed examination?; (5) indicated that *tour of the ship* followed conference?; (6) indicated that *journey or otherworldly journey* followed tour of the ship?; (7) indicated that *theophany* followed journey or other worldly journey?; (8) indicated that *return* followed theophany?; (9) indicated that *aftermath* followed return?; and (10) indicated that the relative order of episodes was maintained" (i.e., capture preceded examination; examination preceded conference, etc.)? (see Bullard, 1998, pp. 6-15).

Although it was not determined how many identical cases are shared between Bullard's earlier work and this research, the 216 cases in this study may still be sufficiently different due to the time lapse separating this research and the earlier results Bullard found. The results in Table 1 parallel earlier findings that revealed a consistency in the way abductions occurred across episodes. That is, the following *percentages* were reported by Bullard: (1) Capture (100); (2) Examination following Capture (70); (3) Conference following Examination (46); (4) Tour following Conference (13); (5) Journey/Otherworldly Journey following Tour (27); (6) Theophany following Journey/Otherworldly Journey (9); (7) Return following

Theophany (66); and (8) Aftermath following Return (71). The difference between percentages then and now are likely affected by differences in the number of cases from sources available to be assessed (see, for example, Jacob's quote about the typical age of abductees who provide data on p. 19). Another factor found in this analysis is that the style of reporting by investigators may have contributed to the perception of how the order of events unfolded. The determination that the sequence of events was maintained in this study was based upon whether a given abduction report mentioned information about a given episode and, thereafter, indicated what happened next. For example, there were no reports encountered during this research that indicated an abductee was first captured, taken on a tour, experienced an otherworldly environment, had a conference about the journey, and then experienced an invasive round of examinations (in cases where such exams took place) before being released/returned. This type of *order in abduction events* was not found. It is also likely that the publication of Bullard's many works and the works of others discussing the field of ufology and abductions may have influenced the order in which some recent reports were written by investigators and, as a result, these factors may or may not have contributed to shaping the order of events. Greatest attention was placed on what abductees said happened. Experiences across abductees that deviated from the sequence of episodes as defined in the Abduction Phenomenon were not evident in the cases reviewed in this research. On this point, Bullard wrote that "out of 103 reports, 84 conform to the standard sequence, and the 19 that deviated from it usually displace no more than one episode" (Bullard, 1998, p.6). Moreover, no investigators whose reports are part of this research indicated in their case write-ups that they were confused by the order of events in which an abductee explained his or her abduction story. What was found in this research was that either information about a particular sequence was in a report, including what preceded or followed it, or it was not. Again, the comparison of percentages indicating consistency in the order of events between the Sequence of Episodes in Bullard's earlier research and the present research is provided in Table 1. The results presented in Table 1 reflect the order of events as *presented in various accounts* in this study and are presented with the recognition of a confounding caveat. Nevertheless, this was an important research question to explore in that Bullard (1998) found that the *sequence of events* is

Table 1: Sequence of Episodes* Analysis Results

Number of Abduction Cases out of 216...	Finding	(Percent)	Bullard's Findings
featuring all eight episodes	1	(0.5)	(81.5)
featuring capture	216	(100)	(100)
featuring examination following capture	155	(71.8)	(70)
featuring conference following examination	99	(45.8)	(46)
featuring tour of the ship following conference	66	(30.5	(13)
featuring journey or otherworldly journey following tour of the ship	12	(5.6)	(27)
featuring theophany following journey/other worldly journey	1	(0.5	(9)
featuring return following theophany	215	(99.5)	(66)
featuring aftermath following return	171	(79.2)	(71)
indicating the relative order among episodes was maintained	216	(100)	(81.5)

Note*: Adapted from Bullard (1998). Abduction Phenomenon. In Jerome Clark (Ed.), The UFO Encyclopedia: The Phenomenon From the Beginning, Vol. 1 (A-K) (2nd Ed.), (pp. 1-26). Detroit: Ruffner.

a key feature of abductions. The results here agree with his previous findings. The caveat borne of apparent differences in methodology noted, the two sets of findings above still appear to reach the same conclusion regarding the order of events in abductions. One observation pertinent to this discussion is that as abduction research continues to grow, perhaps more uniformity in methodology will emerge as a direct consequence of the growth in abduction phenomenon research. Moreover, the only other party reportedly present during an abduction (other than alien beings) has been abductees. In their defense, many ufologists and other clinical or medical professionals maintain that (1) abductees are not any more fantasy prone or neurotic than members of the general public and/or that (2) certain psychopathologies attributed to abductees are not supported by the analysis of data from clinical observations of abductees (Mack, 1994; Appelle, 1996; Lynn and Kirsch, 1996; and Laibow, 2016). The results in Table 1 indicate that the same relative pattern of occurrence of the eight episodes in this research is consistent with results reported in earlier research as mentioned above. That is, among 216 people world-wide, their abduction experiences began and ended in the same order; even though not all abductees experienced every episode or elements within an episode. The answers to Research Question 1 remain an apparently consistent aspect of abduction accounts.

Research Question 2. This question asked: Are all alien groups mentioned in various abduction accounts represented equally? The description of beings given by experiencers became the primary source for arriving at the appearance of alien abductors. Moreover, the various "types of beings" mentioned in Bullard's "Table 3: Description of beings' appearance and behavior" (Bullard, 1998, p. 12) were recorded where mentioned for each assessed case. The use of the *equal frequency hypothesis* (Koenker, 1974; Green, Salkind, & Akey, 2000) enabled the analysis of frequencies (scores) for all abduction reports. Since multiple statistical approaches apply to arriving at a response to Research Question 2, some defense of the approach used here seems appropriate. Thorndike and Dinnel provide an example which is relevant here (2001, p. 374). In their example, 50 percent of a class were education majors, 25 percent were psychology majors, and 25 percent were "other". The results for calculating the Expected N were:

	Observed N	Expected N	Residual
Psychology	20	17.0	3.0
Education	26	34.0	- 8.0
Other	22	17.0	5.0

Test Statistics College Major

Chi Square = 3.882   df = 2   Asymp. Sig = .144.

The above Chi Square value was found to be insignificant and "the data are not inconsistent with the pattern of frequencies specified in the model" (Thorndike and Dinnel, 2001, p. 383). Relative to this research, however, the authors point out that "Of course, there are many other sets of model frequencies, including *equal frequencies for all three groups*, that also "fit the data". We have no way of knowing what "truth" is only that this [i.e., the above calculations, brackets my own] is one possibility" (Thorndike and Dinnel, 2001, pp. 382-383). Lastly, "Chi Square may be used to test a chance, null, or equal frequency hypothesis; or it may be used to test some predetermined or *a priori* hypothesis. An *a priori* hypothesis, as in the above example, is one which is based on previous research or derived from some theory" (Koenker, 1974, p. 106). The choice of an *equal frequency hypothesis,* therefore, is a sound option to address questions of whether or not an event has or has not occurred beyond chance. Moreover, when a mismatch in groupings exist between pre-existing and current data, comparisons would not be meaningful.

Calculations were made to determine if the frequency of descriptions of aliens given by abductees was greater than the frequency of such descriptions expected by chance. The Chi Square result was found to be: $(\chi^2_{(5,\ N=250)} = 475.472$, p = .000 (p < .0005). This result favored the Humanoid species (i.e., grays) as the alien type most responsible for this overall outcome. It is also worth noting that across 216 people in this study, many of these persons saw at least one alien type while other abductees either interacted with or simply saw more than one species during a single abduction event. These observances produced a total of 250 identifications of six alien types by 172 of 216 abductees. The data also show that grays were the species seen most often by abductees (i.e., approximately 68% of the time). These and other findings are presented in Table 2 for Research Question 2.

Research Question 3. This question addressed the calculation of the percentile rank (PR) of alien abduction tendencies. The PR expresses a way of comparing the most active species taking part in human abductions. By implication, PR also makes a statement about the consequences of alien abduction activities with respect to abductees. In other words, the higher the PR, the more experiences people say they had at the hands of these beings.

The total abduction score was used in this research to calculate the PR in two different conditions as reported by abductees: (1) when alien types worked alone (A) and (2) when alien types worked together (T). The data used to calculate the PR for the two work conditions is presented in Table 3, pp. 39-40. PR(s) were calculated separately for the Alone Work Preference Condition(A), (N = 105) and the Together Work Preference Condition (T), (N = 65). Histogram and normal curve graphs of total abduction score frequencies are presented in Figures 1 and 2, pp. 37-38. Abduction case numbers, work condition indications, GPS location values, and Total Abduction scores for 216 cases are presented in Table 4.

One useful benefit of looking at a histogram with a superimposed normal curve is the insight it offers when trying to interpret data in graphical form. Overall, for the alone work preference condition, scores (41.6-73.9) indicate the score range where approximately two thirds of all scores in the distribution are located. For the Together work condition, the range of scores 40.49-76.61 indicates the score range where approximately two thirds of all scores in the distribution are located. The distribution is skewed left due to scores between 40 and 55. The data in the histograms for Figure 1 and Figure 2 reflect the means and standard deviations for the "near" normality of data for the Alone and Together work conditions. Again, Table 2 presents data related to the number and alien types abductees said they saw aboard various craft. As abductions started and eventually ended, abductees indicated that these beings sometimes worked alone and sometimes in a cooperative arrangement (i.e., grays, for example, reportedly received assistance from other types of beings seen aboard various craft). The frequency of scores shown in both Figure 1 (i.e., the **Alone work preference condition)** and Figure 2 (i.e., **the Together work preference condition**) relate only to the species called Humanoids or grays for reasons noted earlier. Notice, however, that this alien type engaged in abductions at a much higher frequency (approximately 12.5) when working alone. The frequency relates to total abduction scores. Compare this frequency of total abduction scores to the much lower one produced when the same species worked cooperatively with other species (approximately 7.5). The pattern in these results support the possibility that Humanoid types are working much harder when alone (by a factor of nearly two) to achieve whatever abduction agenda is at hand. The important point to remember about what is being communicated by all of this "math" is that the numbers and graphs show what abductees have described regarding the nature of their abduction experiences. That is, higher total abduction scores imply that more intensive or invasive abduction experiences are being carried out by one alien type *working alone* more so than when (greys) are working cooperatively with other alien groups.

Table 2: Chi Square Results for Research Question 2: Frequency of Alien Descriptions

Descriptive Statistics

	N	Mean	Std. Dev.	Minimum	Maximum
Alien Description	250	1.9580	1.61360	1.00	6.00

Types*	Observed N	Expected N	Residual
1	170	41.7	128.3
2	17	41.7	-24.7
3	17	41.7	-24.7
4	17	41.7	-24.7
5	10	41.7	-31.7
6	19	41.7	-22.7
Total	250		

Test Statistics

	Alien Description
Chi-Square	475.472
df	5
Asymp. Sig.	.000

Note*: (1-6) = Humanoid, Reptilian/Insectoid being, Nordic, Ordinary Humans, Hybrids, and MORAs (Monsters, Oddities, Robots, and Apes), respectively (Bullard, 1998). Minimum cell frequency is 41.7.

Figure 1: Normal Curve and Histogram of PR Scores for the Alone Condition Work Preference

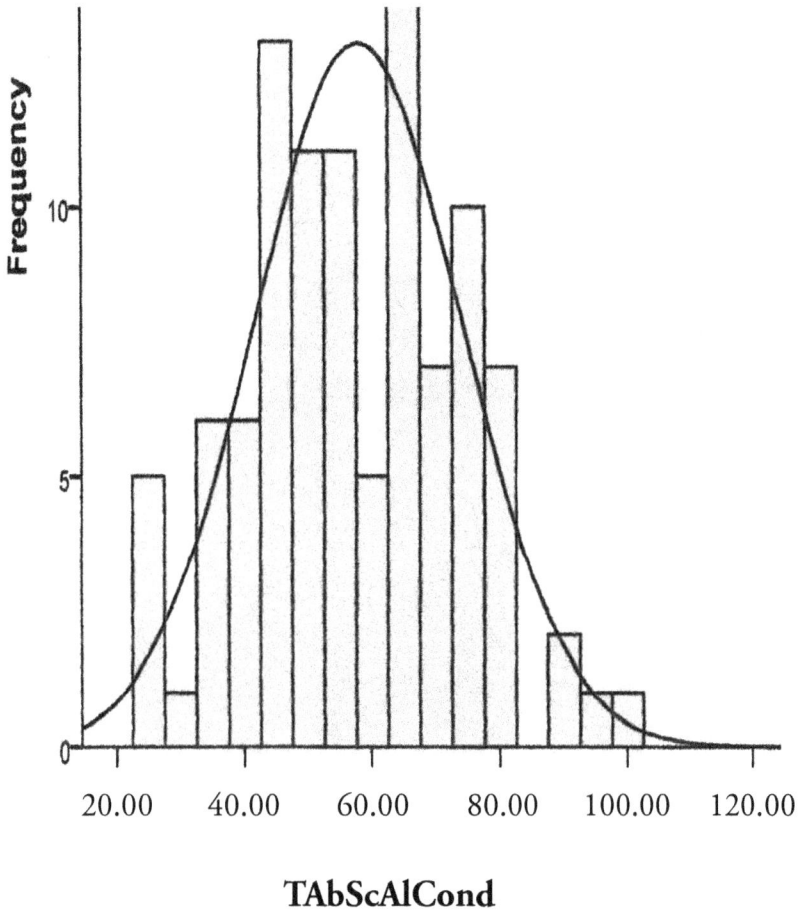

TAbScAlCond

Figure 2: Normal Curve and Histogram of PR Scores for the Together Condition Work Preference

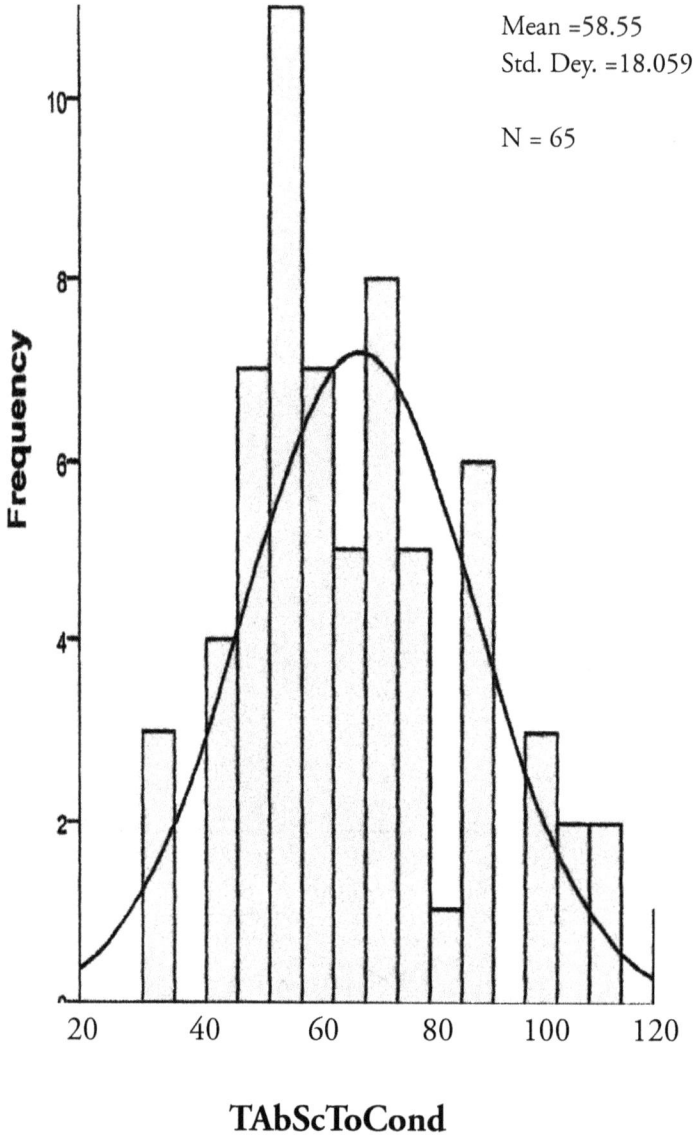

Mean =58.55
Std. Dey. =18.059

N = 65

TAbScToCond

Table 3: [PR] Results for Alone (A) and Together (T) Work Preference Conditions and Total Abduction Scores

[29.30] (A) 49 [84.20] (A) 74	[2.45] (A) 23	[2.45] (A) 26
[23.6] (A) 46 [82.66] (A) 73	[29.30] (A) 49	[48.03] (A) 57
[64.94] (A) 64 [62.62] (A) 63	[25.19] (A) 47	[29.30 (A) 49
[19.45] (T) 43 [53.19] (T) 60	[22.65] (T) 45	[35.98] (A) 52
[13.53] (A) 40 [23.26] (A) 46	[19.45] (T) 43	[87.00] (A) 76
[71.45] (T) 69 [17.98] (A) 43	[23.26] (A) 46	[87.00] (A) 76
[24.35] (T) 46 [93.57] (T) 86	[57.57] (T) 62	[62.62] (A) 63
[44.36] (T) 56 [99.55] (A) 100	[57.57] (T) 62	[11.63] (T) 37
[25.19] (A) 47 [13.53] (A) 40	[37.92] (T) 53	[19.45] (T) 43
[35.98] (A) 52 [7.91] (A) 35	[35.83] (T) 52	[69.95] (T) 68
[43.12] (A) 55 [59.72] (T) 63	[59.72] (T) 63	[71.85 (T) 69
[9.90] (A) 37 [79.3] (A) 71	[29.84] (T) 49	[21.41] (A) 45
[33.79] (T) 51 [79.3] (A) 71	[67.21] (A) 65	[22.65] (T) 45
[85.92] (T) 78 [4.53] (T) 28	[64.94] (A) 64	[29.84] (T) 49
[9.90] (A) 37 [23.26] (A) 46	[2.45] (A) 26	[77.49] (A)70
[53.19] (T) 60 [2.12] (A) 25	[43.12] (A) 55	[38.32] (A) 70
[92.45] (A) 81 [67.21] (A) 65	[67.21] (A) 65	[90.53](A) 79
[31.46] (A) 50 [9.61] (T) 35	[62.62] (A) 63	[13.53] (A) 40
[38.32] (A) 53 [40.70] (A) 54	[97.52] (T) 94	[87.0] (A) 76
[64.94] (A) 64 [99.63] (T) 107	[84.2] (A) 74	[50.49] (A) 58
[7.04] (A) 34 [85.65] (A) 75	[69.41] (A) 66	[93.29] (A) 82
[4.53] (T) 28 [71.55] (A) 67	[29.84 (T) 49	[84.65] (T)77
[37.92] (T) 53 [52.96] (A) 59	[42.20] (T) 55	[22.65] (T) 45
[24.35] (T) 46 [81.88] (T) 75	[85.92] (T) 78	[25.19] (A) 47
[26.12] (T) 47 [31.46] (A) 50	[16.41] (A) 42	[46.57] (T)57
[69.41(A) 66 [12.75] (T) 38	[92.85] (T) 85	[85.92] (T)78
[50.49] (A) 58 [25.19] (A) 47	[45.56] (A) 56	[27.21] (A) 48
[21.01] (T) 44 [97.93] (A) 89	[50.49] (A) 58	[91.53] (A) 80
[57.85] (A) 61 [96.92] (A) 88	[35.83] (T) 52	[11.02] (A) 38
[21.41] (A) 45 [21.41] (A) 45	[45.56] (A) 56	[10.59] (T) 36
[45.56] (A) 56 [73.69] (T) 70	[19.45] (T) 43	[12.23] (A) 39

Table 3 (Work Preference Conditions and Total abduction Scores (continued)

[68.00] (T) 67	[69.41] A) 66	[95.92] (T) 90	[29.30] (A) 49
[9.90] (A) 37	[98.93] (A) 95	[93.29] (A) 82	[27.95] (T) 48
[46.57] (T) 57	[91.53] (A) 80	[45.56] (A) 56	[40.05] (T) 54
[33.79] (T) 51	[57.57] (T) 62	[26.12] (T) 47	[59.72] (T) 63
[3.73] (A) 29	[9.90] (A) 37	[62.62] (A) 63	[35.38] (A) 52
[98.55] (T) 98	[67.21] (A) 65	[81.88] (T) 75	[73.61] (A) 68
[98.34] (T) 97	[67.21] (A) 65	[94.24] (T) 87	[31.46] (A) 50
[69.41] (A) 66	[4.03] (T) 27	[79.30] (A) 71	[88.26] (A) 77
[50.99] (T) 59	[65.99] (T) 66	[81.03] (A) 72	[17.97] (T) 42
[62.62] (A) 63	[69.41] (A) 66	[82.66] (A) 73	[92.45] (A)81
[21.01] (T) 44	[48.03] (A) 57	[82.66] (A) 73	[2.45] (A)26
[40.70] (A) 54	[85.65] (A) 75		

Research Question 4 and Hypotheses. Research Question 4 asked: What are the insights that might be gained from examining variables such as *age, abduction scores, alien work conditions, and modes of communication?* Four hypotheses were developed to explore Research Question 4:

Hypothesis 1, Research Question 4 postulated that there will be no significant difference in the observed and expected frequencies of men and women in three age groups abducted world-wide compared to what would be expected by chance. The overall result indicated that the number of men and women abducted across three age groups was significantly different from what would be expected by chance ($\chi^2_{(2, N=216)}$ = 68.778, p = .000 (p < .0005). Gender for the three groups was also significant. Across the two largest age groups (≤30 and 31-49), the follow-up analysis indicated that the frequencies were greater than what would be expected by chance and the significantly different outcome favored the ≤30 age group ($\chi^2_{(1, 198)}$ = 5.838, p = .016). The ≤50 age group was eliminated, since its Expected N was larger than its Observed N. With regard to gender, the number of men compared to women abducted world-wide was not significantly different from what would be expected by chance ($\chi^2_{(1, 198)}$= 2.909, p = .088). The results for the two analyses are shown in Tables 5 and 6.

Hypothesis 2, Research Question 4. This hypothesis investigated if a significant difference existed in the frequency of Total Abduction scores sorted by (1) alien work condition preference, (2) global geographical region, (3) ranges of Total Abduction Scores, and (4) ranges of derived GPS locations for previous abductions compared to what would be expected by chance? The data used for these analyses, as mentioned earlier, pertained to the Humanoid species only. The results indicated that for each of the four equal frequency hypotheses tested, all four were significantly different from the frequency expected by chance. That is, (1) humanoids worked significantly more in the alone condition than in the together condition with other alien species when abducting humans ($\chi^2_{(1, 170)}$ = 9.412, p = .002); (2) Humanoids visited the North Latitude, West Longitude Region of this planet significantly more times than any of the other three geographical regions when abducting humans ($\chi^2_{(3, 170)}$ =159.506, p = .000 (p < .0005). The North latitude, West longitude region is the *region of locations* favored by this result; (3) the (48-67) range of abduction scores was significantly greater than any of the other ranges of abduction scores $\chi^2_{(3, 170)}$= 53.200, p = .000 (p < .0005); the result exceeded the number expected by chance; and (4) the frequency for the GPS range of values (3.1-7.0) exceeds the frequency expected by chance ($\chi^2_{(2, 170)}$ = 122.871, p = .000 (p < .0005). All other ranges of GPS values were eliminated since their Observed N's were less than their Expected N's. The results for Hypothesis 2 are shown in Table 7. Additionally, with regard to finding (4) above, readers should note that in the *reconversion* back to conventional latitude and longitude information, **the 3** in (3.1) is segment 3 where GPS latitudes range from (31-45)° north of the equator and **the 1** in (3.1) is segment 1 where GPS longitudes range from (0-15)° west of the Prime Meridian. Similarly, 7 in (7.0) is segment 7 where GPS latitudes range from (0-15)° south of the equator and 0 in (7.0) is the Prime Meridian, or the start of longitudes west (or east) of the Prime Meridian. The two corridors of abduction locations are shown in Figure 3, p. 60. It is interesting that the procedure used here produced results that agree with previous knowledge about geographical locations of alleged alien activity.

A review of GPS location data in Table 3, p. 29 indicates examples of cities within the two corridors in Figure 3, p. 48 (i.e., cities in the ranges (3.1-7.0).

Lastly, the highly significant correlation between GPS Range of Values and Region was very strong (r=.668) and highly significant (p = .000 or p < .0005). Most interesting is that the frequency for the range of GPS values (3.1-7.0) was greater than what would be expected by chance (see GPS Range of Values (Locations) within Table 7). Although weak and inverse, (r = -.291), the correlation between GPS Range of Values and Score Range was highly significant (p=.000 or p < .0005). With Alaska as a reference point, the (3.1) abduction corridor encircles the globe and passes through most of America, Europe, The Middle East, and China, while the (7.0) corridor passes through most of Mexico, Northern/Central Africa, Saudi Arabia, Southern China, and most of India as it encircles the globe. Note as well the supporting results on p. 48 and pp. 98-100.

Hypothesis 3, Research Question 4. This hypothesis postulated that there would be no significant correlation between (1) the frequency of alien work condition preferences, (2) the frequency of ranges of abduction phenomenon total scores, and (3) the four geographical regions as predictor variables and the frequency of ranges of GPS derived abduction locations **as the dependent variable.** Standard Multiple Regression results for the correlation between GPS range of values (i.e., where abductions have reportedly occurred) as the dependent variable and work condition, geographical region of abduction activity, and range of total abduction scores as the predictor variables indicate that the regression equation with all three predictors was significantly related to the GPS range of values R = .687, R^2 = .471, Adjusted R^2 = .462); $F_{(3, 166)}$ = 49.340, p = .000 (p <.0005). No collinearity issues of concern (Tolerance > .10 and VIF < 10). The correlation data is shown in Table 8. The prediction equation associated with the correlation result (unstandardized weights) is shown below:

$Z_{\text{GPS Range of Values}}$ **= -.037Z** $_{\text{Work Condition}}$ **+ .480ZRegion - .093ZAbduction Range of Values + 1.337.** The maximum Mahal. Distance = 9.340; critical value = 16.27 and no cases (outliers) > 1 relative to Cook's distance (.060).

In a follow-up Standard Multiple Regression analysis for possible model changes due to differences in strength of the independent predictor variables, the result indicated that Region and Total Abduction Score accounted for 46.4 percent of the variance in the prediction of GPS Range of Values. Work condition did not contribute very much toward changing R^2 (F (1,166) = .386, p= .535) with an R^2 change = .001. Hence, although present in the prediction equation above, work condition can be omitted from the calculation of predicted values for (new) abduction locations. The import of this result is that researchers assessing abduction cases in the manner done here and desiring to focus on locations specific to future abductions may wish to consider where some alien species, Humanoids, particularly, may likely visit this planet.

Additional findings from the above analysis indicate that geographical *Region of abduction* made the strongest unique contribution (standardized beta = .657) to explaining GPS Range of Values and the contribution was statistically very highly significant (p =.000 or p< .0005). *Score Range* (standardized beta = -.153), or the abduction total score/indication of treatment during an abduction, made the second strongest unique contribution to explaining GPS Range of Values and the contribution was statistically highly significant (p = .008). *Work Condition* preference (standardized beta = -.035) made the weakest unique contribution of all predictors of GPS Range of Values. The contribution also was not statistically significant (p = .535). Moreover, the correlation between GPS Range of Values and Region (r = .668) was strong and very highly significant (p =.000 or p< .0005) and the correlation between GPS Range of Values and Score Range (r = -.291) was weak, inverse, and highly significant (p = .002). Lastly, the correlation between GPS Range of Values and (alien) Work Condition preference (r = .027) was weak and not statistically significant (p = .365). This effectively reduces the prediction equation to:

G. P. S. Range of Values = .480 x (Region Number...w/values between 1-4) + 1.337. The resulting G. P. S. Range of Values could range from (2.0-4.0), (4.1-6.1), (6.2-8.2), or (8.3-15.0).

Hypothesis 4, Research Question 4. This hypothesis investigated if there was a significant difference in reported modes of communication (i.e., telepathy, no communication, and other [i.e., grunts, gestures, alien language, use of English]) used by alien beings while working alone or working with other beings during abduction events. The overall Chi-Square result indicated that there was a significant difference in the styles of communication used by alien beings while working alone or working together with other beings during abduction events ($\chi^2_{(2, N=184)}$ = 76.967, p=.000 (p = .0005). A follow-up Chi-Square test determined that a significant difference was found in the frequency of telepathy reported versus the frequency of "no communication" reported while different beings worked either alone or together with other beings ($\chi^2_{(1, N=175)}$ = 7.000, p = .008). The outcome favored the use of telepathy in the alone condition as the style of communication used by Humanoids. Readers will note that the Expected N for the third style of communication (grunts, etc.) was eliminated from the follow-up test, since its Expected N was greater than its Observed N in the initial test of frequencies. This left the alone condition with telepathy and "no communication" categories equal to frequencies of 66 and 49, respectively, and the together condition with telepathy and "no communication" categories equal to frequencies of 39 and 21, respectively. These results indicate that telepathy occurred at a frequency greater than what would be expected by chance. The results for the two tests are shown in Table 9.

The next area of inquiry focused on brief descriptive information about the appearance of alleged abductors so that, perhaps, what abductees say they experienced might be shared. The reported appearance of the creatures listed at the bottom of Table 2, p. 26, are partial descriptions adapted from accounts by abductees and, in part, from (1) The Field Guide to Extraterrestrials (1996) by Patrick Huyghe, with illustrations by Harry Trumbore and (2) Bullard's Abduction Phenomenon (1998):

Type	Description
Humanoids	Popular image of the extraterrestrial; short humanlike beings with large bulbous heads, hair-less bodies; almond-shaped wrap-around eyes, black eyes but no pupil; skinny arms; fragile bodies; gray skin; (short variety) 4.5 ft to seven ft tall (giant variety); some may have turtle-like lower half of face.

Reptilian/Insectoids | Green, scaly entities; large stature; muscular build; swamp-type creature; faces with oval eyes; Insect-like beings with large insect-like eyes; spindly arms; grasshopper/praying mantis appearance also in this group; 7-8 ft tall.

Nordic | Blond hair; bright blue eyes; white to nearly translucent skin; Scandinavian appearance in many accounts; beautiful; 6.5 ft tall; humanoid variant; other variants in this class; translucent skin also described for this class.

Ordinary Humans | To be distinguished from other abductees (seen in tubes); this group on board craft assisting grays and Nordics; this group could "pass for the guy on the street"; speaks English; seldom seen alone operating a craft; 4-5 ft tall; some may only communicate with grunts or growls as in the Villas- Boas case; long, slanted eyes; blue in color; high cheek bones; wide face.

Hybrids | Thought to be mix between human beings and alien DNA; stringy hair; pale white skin; some seen in early after birth stages; others older; short, 4-4.5 ft tall; most seen aboard craft were women caring for human-looking children.

Monsters, Oddities, Robots & Apes | As the names imply, some are large or small body types; can be covered in hair, ape-like or even animal looking with disproportionately long arms; would be described as monsters by human standards; others have tin can looking bodies; actually part or all mechanical; some with no visible eyes, but can have multiple eyes or instruments serving a communication function; others are clearly "odd" in that their bodies do not appear human at all; some are translucent; others jelly like or even have shiny, metallic skin with two arms; heights vary from 3-5 ft and taller.

Percentile Rank and Alleged Abducting Groups

It is evident from the literature that it is not unusual for mixtures of alien beings to be identified in connection with different abduction cases in different locations or even in the same location, but involve different people or the same people at different times (multiple abductees, for example). This type of behavior among alien beings has been reported all over the world.

The percentile rank (PR) (or percentile score) "indicates a student's relative position in a group in terms of the percentage of students scoring lower" (Linn and Gronlund, 1995, p. 448). Without stretching this psychometric term's meaning or interpretation too far, consider the following: one object of learning and testing (among humans) is to get all test items correct or to demonstrate mastery of content by making the highest score possible. One's overall performance through learning and testing occurring many, many times may, indeed, help one reach such a desired goal. In many ways, the abduction experience has attributes akin to how human learners might seek to improve their relative standing within a group. That is, in a very real sense, the "performance" during abduction episodes of alien beings, at least from descriptions given by abductees, seems to imply that these beings have not quite mastered an understanding of human beings *as the content or test material* alien beings are interested in mastering. In case after case of reported abductions world-wide, the reported behavior of these beings seems to imply that alien groups are attempting to increase their knowledge of humans (i.e., earn a high PR) by abducting them, then exposing them to different degrees of invasive testing and psychological trauma without a care. This behavior in the pursuit of an ultimate goal which remains a matter of reasonable uncertainty (Jacobs, 1992). In this book, PR is a suitable way to express how intensely alien beings are reportedly working alone and together to achieve whatever their objectives may be. If abductions are attempts to understand humans, biologically or socially, then the higher the PR, the more that understanding is or has progressed.

In education and measurement circles, PR conveys two pieces of information simultaneously. A score that gets transformed into a percentile rank not only says how an individual is performing relative to a group, but also says how others in the same group are performing as well. When applied to abductions, the PR metric provides not only valuable information about the groups of beings conducting abductions, but such a metric also conveys a sense of the correspondingly negative impact persons both experience and attempt to understand *following the aftermath* of an abduction event. As the number of reported invasive procedures increase, the PR increases as well. This hidden psychological impact is often proportional to the abduction scores reported in this study. The consistency of abduction narratives appears to support this numerical interpretation of PR (and its psychological toll) experienced across individual abduction events. Results for the calculation of percentile rank followed the procedure described by SPSS (Green, Salkind, & Akey, 2001). Close agreement with SPSS PR values were also found using the procedure by Salvia, Yesseldyke, & Bolt (2007).

Global Positioning System (GPS) Data Reduction Procedure

To facilitate analyses featured in this study, a procedure was devised to represent, as one number, the latitude and longitude information identifying the location of an abduction event. An example will be instructive as to how GPS data was transformed. Note that Itaperuna, Brazil, for example, is found at latitude S 20°54'27.837" and longitude W42°1'37.998". To achieve the transformation to one number, notice that *latitude* can be divided into twelve-15° segments as follows:

(0-15)°	(16-30)°	(31-45)°	(46-60)°	(61-75)°	and	(76-90)°
1	2	3	4	5		6

for locations **North of the equator (numbers 1-6)** and **South of the equator (numbers 7-12)**, respectively:

(0-15)°	(16-30)°	(31-45)°	(46-60)°	(61-75)°	and	(76-90)°.
In 7	8	9	10	11		12

a similar fashion, degrees of *longitude* **East of the Prime Meridian** can be divided into twelve-15° segments (i.e., **numbers 13-24**) as follows:

(0-15)°	(16-30)°	(31-45)°	(46-60)°	(61-75)°	(76-90)°	(91-105)°
13	14	15	16	17	18	19

(106-120)°	(121-135)°	(136-150)°	(151-165)°	(166-180)°.
20	21	22	23	24

For locations **West of the Prime Meridian**, twelve 15° segments (i.e., **numbers 25-36)** can be formed as follows:

(0-15)°	(16-30)°	(31-45)°	(46-60)°	(61-75)°	(76-90)°	(91-105)°	(106-120)°	(121-135)°
25	26	27	28	29	30	31	32	33

(136-150)	(151-165)	(166-180)
34	35	36

The final transformation is achieved by finding the proper direction, range of degrees, and the segment number as outlined above. For the example above, the *southern* latitude can be found in segment (**8**) and the *western* longitude can be found in segment (**27**). Thus, the transformation results in the number **8.27.** Transformation numbers were calculated for all G. P. S. locations of abduction cases included in this study. Readers should note that the transformation of latitude and longitude information into a single number was accomplished by taking advantage of the fact that latitude and longitude are divided into 15° segments, north and south of the equator and east and west of the Prime Meridian. It was desired in this research to transform the latitude and longitude information associated with abduction sites into *one number* in order to correlate geographical locations with other variables.

Table 4. Work Condition* (A/T), [Case Number], GPS** Location Value, and (Total Abduction Scores) of Selected Cases by Geographical Region

North Latitude East Longitude	North Latitude West Longitude	South Latitude East Longitude	South Latitude West Longitude
(A) [9] 4.13 (49)	(A)[1] 3.29 (81)	(T) [13] 9.23(28)	(T) [7]8.28 (53)
(A) [10]4.13 (46)	(A) [2] 3.29 (75)	(A) [38] 9.22(46)	(T) [8] 8.28(52)
(T) [15] 4.14 (43)	(A) [4] 3.29 (73)	(A) [67] 9.22 (25)	(T) [23] 8.28 (63)
(A) [16] 3.22 (40)	(A) [5] 3.29 (63)	(*) [76] 9.22 (26)	(T) [30] 8.27 (49)
(T) [17] 3.17 (69)	(T) [6] 4.25 (60)	(A) [85] 8.33 (65)	(A) [46] 9.29 (65)
(T) [27] 3.18 (46)	(A) [11] 4.25 (46)	(A) [90] 9.22 (50)	(*) [57] 9.29 (27)
(A) [34] 3.23 (42)	(A) [12] 4.25 (43)	(T) [99] 9.23 (35)	(*) [60] 7.27 (17)
(A) [39] 3.23 (47)	(*) [18] 3.29 (58)	(A) [112] 9.22 (23)	(A) [122] 7.28 (64)
(A) [53] 3.15 (52)	(T) [20] 2.31 (86)	(A) [114] 8.23 (49)	(*) [124] 9.29 (24)
(A) [69] 3.21 (55)	(A) [21] 3.30 (100)	(A) [115] 8.23 (47)	(*) [130] 9.28 (55)
(A) [86] 4.15 (37)	(A) [22] 3.32 (40)	(T) [120] 8.22 (45)	(*) [142] 9.29 (22)
(T) [65] 3.13 (51)	(A) [25] 4.25 (35)	(T) [126] 8.14 (43)	(A) [154] 8.27 (26)
(T) [89] 2.17 (78)	(T) [26] 3.30 (63)	(A) [134] 9.23 (46)	(*) [158] 8.28 (24)
(A) [94] 4.14 (37)	(A) [28] 3.29 (71)	(T) [135] 9.23 (62)	(*) [167] 8.27 (67)
(T) [113] 4.13 (60)	(A) [31] 2.29 (71)	(T) [137] 9.23 (62)	(A) [175] 8.28 (55)
(A) [119] 5.14 (65)	(T) [29] 3.29 (37)	(A) [139] 9.23 (64)	(A) [193] 8.27 (61)
(T) [113] 4.13 (60)	(T) [113] 4.13 (60)	(A) [31] 2.29 (71)	(T) [137] 9.23 (62)
(A) [175] 8.28 (55)	(A) [119] 5.14 (65)	(T) [29] 3.29 (37)	(A) [139] 9.23 (64)
(A) [193] 8.27 (61)	(T) [113] 4.13 (60)	(A) [14]4.13(64)	(A) [3]3.29 (74)
(*) [55] 8.21(25)	(*)[19] 9.29 (17)		
(A) [31] 2.29 (71)	(T) [137] 9.23 (62)	(A) [175] 8.28 (55)	
(A) [119] 5.14 (65)	(T) [29] 3.29 (37)	(A) [139] 9.23 (64)	(A) [193] 8.27 (61)
(A) [24] 3.17 (63)	(T) [32] 4.25 (43)	(*) [149] 9.22 (37)	(A) [212] 9.22 (58)
(A) [192] 3.15 (49)	(A) (40) 3.32 (45)	(T) [180] 9.23 (46)	(A) [24] 3.17 (63)
(T) [32] 4.25 (43)	(*) [149] 9.22 (37)	(A) [212] 9.22 (58)	(A) [168) 4.13 (26)
(T) [33] 3.29 (68)	(A) [161] 8.22 (34)	(*) [177} 4.14 (25)	(*) [36] 3.31 (20)
(T) [162] 8.22 (28)	(A) [191] 5.14 (57)	(T) [37] 2.30 (69)	(T) [178] 9.23 (53)
(A) [192] 3.15 (49)	(A) (40) 3.32 (45)	(T) [180] 9.23 (46)	(A) [197] 4.13 (52)
(T) (41) 2.30 (45)	(T) [181] 9.23 (47)	(A) [206] 4.14 (76)	(T) (42) 1.30 (49)
(A) [190] 9.22 (66)	(A) [207] 4.14 (76)	(*) (13) 3.32 (28)	(*) [202] 9.22 (21)
(*) [216] 3.13 (61)	(A) (44) 3.30 (70)	(T) [136] 9.23 (44)	(A) [148] 3.17(63)
(*) [45] 3.29 (28)	(A) [48] 3.29 (70)		

Table 4. Total Abduction Scores and Data for Selected Cases (continued)

North Latitude East Longitude	North Latitude West Longitude	South Latitude East Longitude	South Latitude West Longitude
(A)[56]3.30(45)	(T)[82] 3.33(75)	(*)[106] 3.30 (27)	(T) [140] 5.33(85)
(A) [59]4.25(56)	(A) [83 3.31(50)	(A) [107]3.29(65)	(A)[141]3.29 (56)
(T) [58] 3.29(67)	(T) [84] 3.32 (38)	(A) [108) 4.33(65)	(A) [143] 3.31 (58)
(A) [61] 3.25 37)	(A) [87] 3.29 (47)	(T) [109]3.29 (27)	(*) [144] 3.31 (19)
(*) [62] 3.32 (30)	(*) [88] 3.32 (35)	(*) [110] 2.31 (25)	(T) [145] 3.30 (52)
(T) [63] 3.29 (57	(A) [91] 3.31(89)	(T) [111] 4.31(66)	(*) [146] 3.29 (41)
(T) [66] 4.25 (51)	(A) [92] 3.29 (88)	(*) [116]3.20 (21)	(*) [147] 4.30 (21)
(A) [68] 2.30 (29)	(A) [93] 4.25 (45)	(*) [117] 3.29 (14)	(A)[150]3.32 (56)
(T)[70]3.32 (98)	(T) [95] 2.31 (70)	(T) [118] 3.30 (94)	(T) [151] 3.30 (43)
(T) [71] 3.32 (97)	(*) [96] 4.25 (24)	(*) [121]3.30 (14)	(T) [152] 3.29 (90)
(A)[72] 4.31 (66)	(A)[97]3.30 (66)	(A)[123]3.31(74)	(A) [153] 3.29 (82)
(T) [73] 3.25 (59)	(A) [98] 3.30 (95)	(*) [128] 3.29 (28)	(A) [155] 2.30(56)
(A) [74] 3.32 (54)	(*) [100] 3.31 (34)	(A) [127]3.29(66)	(*) [156]3.32(33)
(T) [77] 3.29(107)	(*) [101] 3.31 (34)	(*) [129] 3.30 (23)	(T) [157] 2.29(47)
(A) [78] 3.29 (75)	(*) [102] 3.31(30)	(T) [131] 3.25 (49)	(A) [159] 3.30(63)
(A) [79] 3.29 (67)	(A) [103] 3.30 (80)	(T) [132]3.30 (55)	(*) [160] 3.33 (41)
(*) [80] 3.32 (39)	(T) [104]3.33 (62)	(T) [133] 3.32(78)	(*)[163] 2.31(40)
(A) [81] 2.30 (59)	(A)[105]3.29 (37)	(A)[138]2.30(42)	(T) [164] 3.32(75)
(T) [165] 3.29(87)	(*) [166] 3.29(54)	(T) [169]3.31(48)	(A) [170]3.30(44)
(A) [171] 3.29 (71)	(*) [172] 4.32 (55)	(T)[173] 3.29 (54)	(A)[174]3.31(81)
(A) [176]3.32 (72)	(T)[179] 3.30 (57)	(T) [182]4.31(63)	(T) [183] 3.30 (44)
(A) [184] 3.30 (76)	(T) [185] 3.31 (78)	(A) [186]3.32 (52)	(A)[187] 3.29 (57)
(T)[208]3.32 (77)	(T) [209] 3.30 (36)	(T) [210] 3.30 (42)	(A) [211] 3.29 (26)
(T) [213] 3.31 (45)	(A) [214] 3.32 (39)	(A) [215] 3.32 (63)	(A) [35] 3.30 (54)
(A) [64] 3.29 (47)	(A) [75] 3.30 (49)	(A) [125] 4.25 (66)	(A) [188] 3.29 (58)
(A) [189] 3.30 (48)	(A) [194] 1.30 (68)	(*) [195] 3.29 (33)	(*) [196] 3.33 (30)
(A) [198] 3.32 (80)	(A) [199] 3.32 (50)	(A) [200] 3.30 (73)	(A) [201] 3.30 (82)
(A) [203] 3.32 (38)	(A) [204] 3.29 (77)	(*) [205] 3.29 (16)	(*) [47] 4.29 (37)
(A) [49] 3.30 (79)	(*) [50] 4.25 (30)	(*)[51] 3.30 (70)	(A) [52] 2.30 (40)
(A) [54] 3.29 (53)	(*) [43] 3.22 (20)		

Note: (*) = Not a Humanoid (Alone/Together) work condition case (**) = (derived) Global Positioning System value follows [Case Number]

Steps taken to replace familiar longitude and latitude numbers with a single decimal number greatly facilitated correlations performed in this study. In the example above, if one wished to reverse the process and go back to the original latitude and longitude values, one need only remember that segment 8 in [8.27] takes one back to the range of degrees, (16-30)°, associated with the original latitude value of S20°54'27.837". Similarly, note that segment 27 (the decimal part of 8.27) is located in the range of degrees (31-45) of the original longitude, W42°1'37.998".

In this process, the derived values (i.e., proxies for latitude and longitude) are used when calculations are done and *not the original latitude and longitude values*. Most importantly, note that the result of using these derived numbers produced outcomes that agree with what is already known about, in this case, an area on the globe where an abundance of alien activity is alleged to have occurred (see also Table 7; Hypothesis 2, Research Question 4, p. 31). Data in Table 4, preceded by an asterisk in parenthesis, were not associated with *Humanoid activity* but with other much lesser represented alien types. These cases were omitted from statistical analyses for this reason, even though derived values were calculated for location data of each case. Otherwise, all transformation numbers as shown in Table 4 along with Total Abduction scores by geographical region are associated with *Humanoid activity only*.

Table 5: Chi-Square Results for Men and Women Abducted Across Three Age Groups Descriptive Statistics

	N	Mean	Std. Dev.	Minimum	Maximum
Age Groups	216	1.5463	.64533	1.00	3.00
Gender	216	1.4120	.49335	1.00	3.00

Age Groups	Observed N	Expected N	Residual
1(= ≤30)	116	72.0	44.0
2(=31-49)	82	72.0	10.0
3(= ≥50)	18	72.0	-54.0
Total	216		

Gender			
	Observed N	Expected N	Residual
1 (=Men)	127	108.0	19.0
2 (=Women)	89	108.0	-19.0
Total	216		

Test Statistics		
	Age Groups	Gender
Chi Square	68.778	6.685
df	2	1
Asymp. Sig.	.000	.010

Table 6: Follow-up Chi-Square Analysis for Age Groups and Gender Abductions

Descriptive Statistics					
N	Mean	Std. Dev.	Minimum	Maximum	
Age Groups	198	1.4141	.49382	1.00	2.00
Gender	198	1.4394	.49757	1.00	2.00

Age Groups	Observed N	Expected N	Residual
1 (= ≤ 30)	116	99.0	17.0
2 (= (31-49)	82	99.0	-17.0
Total	198		

Gender			
	Observed N	Expected N	Residual
1 (= Men)	111	99.0	12.0
2 (= Women)	87	99.0	-12.0
Total	198		

Test Statistics		
	Age Groups	Gender
Chi-Square	5.838	2.909
df	1	1
Asymp. Sig.	.016	.088

Table 7: Chi-Square Results for Four Equal Frequency Test Conditions

Descriptive Statistics					
	N	Mean	Std. Dev.	Minimum	Maximum
WorkCond	170	1.3824	.48740	1.00	2.00
Region	170	2.0941	.69866	1.00	4.00
Score Range	170	1.9941	.83911	1.00	4.00
GPSRangeVal	170	2.1059	.51080	1.00	3.00

Work Condition			
	Observed N	Expected N	Residual
1(=Alone)	105	85.0	20.0
2(=Together Cond)	65	85.0	-20.0
Total	170		

Region			
	Observed N	Expected N	Residual
1 (=NlatElong)	25	42.5	-17.5
2 (=NlatWlong)	113	42.5	70.5
3 (=SlatElong)	23	42.5	-19.5
4 (=SlatWlong)	9	42.5	-33.5
Total	170		

Table 7 Chi-Square Results for Four Equal Frequency Test Conditions (continued)

Score Range			
	Observed N	Expected N	Residual
1 (=26-47...Very Low)	53	42.5	10.5
2 (=48-67)...Low)	72	42.5	29.5
3 (=68-89...Medium)	38	42.5	-4.5
4 (=90-108...High)	7	42.5	-35.5
Total	170		

GPS Range of Values (Locations)			
	Observed N	Expected N	Residual
1 (=0-3.0...Low)	14	56.7	-42.7
2 (=3.1-7.0...Med)	124	56.7	67.3
3 (=7.1-9.5...High)	32	56.7	-24.7
Total	170		

Test Statistics				
	WorkCond	Region	ScoreRange	GPS Range of Values
Chi-Square	9.412	159.506	53.200	122.871
df	1	3	3	2
Asymp. Sig.	.002	.000	.000	.000

Table 8: Standard Multiple Regression Results for Selected Variables

Descriptive Statistics			
	Mean	Std Deviation	N
GPS Range of Values	2.1059	.510880	170
WorkCond	1.3824	.48740	170
Region	2.0941	.69866	170
Score Range	1.9941	.83911	170

Correlations				
	GPS Vals	WorkCond	Region	Score Range
Pearson Correlations GPSRange of Values				
WorkCond	.027			
Region	.668	.102		
ScoreRange	-.291	.034	-.100	
Sig. (1-tailed) GPSRange of Values		.365	.000	.002
WorkCond	.365		.092	.328
Region	.000	.092		.097
ScoreRange	.000	.328	.097	
N	All Variables	170	170	170	170

Model Summary				
1	R	R Square	Adjusted R Square	Std. Error of the Estimate
	.687	.471	.462	.37472

Model	Unstandardized Coefficients		Standardized Coefficients	Correlations				
	B	Std. Error	Beta	t	Sig.	Zero-Order	Partial	Part
1 (constant)	1.337	.139		9.662	.000			
Work Cond	-.037	.060	-.035	-.662	.535	.027	.048	-.035
Region	.480	.042	.657	11.512	.000	.668	.666	.650
Score Range	.093	.035	-.153	-2.687	.008	-.219	-.204	-.152

Table 9: Chi-Square Results for Three Equal Frequency Test Conditions Three Styles of Communication

	Observed N	Expected N	Residual
1 (=Telepathy)	105	61.3	43.7
2 (=No Communication)	70	61.3	8.7
3 (=Other; grunts, English, etc.)	9	61.3	-52.3
Total	184		

Test Statistics	
Two Styles of Communication	
Chi-Square	76.967
df	2
Asymp. Sig.	.000

	Observed N	Expected N	Residual
1(=Telepathy)	105	87.5	17.5
2(=No communication)	70	87.5	-17.5
Total	175		

Test Statistics					
Style					
Chi-Square	7.000	df	1	Asymp. Sig.	.008

Analysis of Close Encounters of the Fourth Kind

Figure 3.

Regions of GPS Locations Favored in Previous Abductions

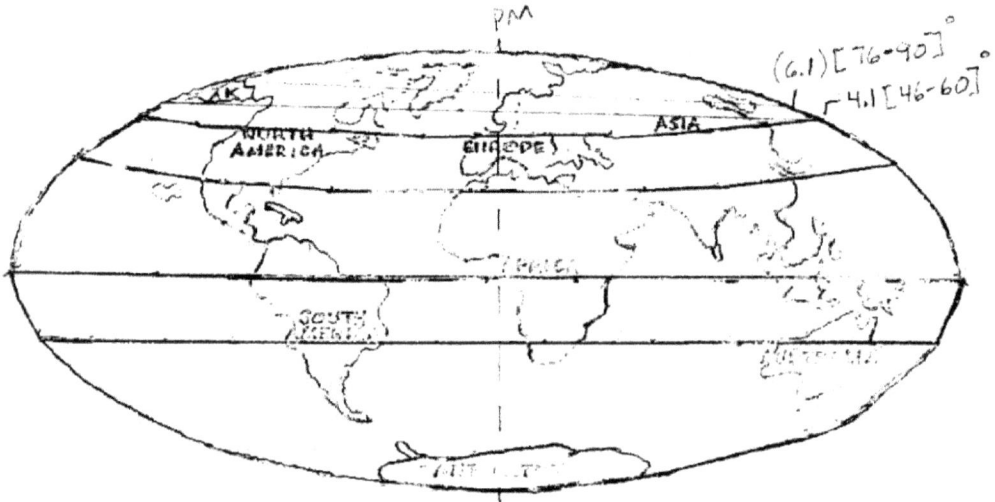

Figure 3. Represents the significant results for favored locations involving previous abductions (p = .000 (p < .0005) for both region of the world and (derived) GPS range of values. The world map is courtesy of litip://IA v,, fro:. ιι oridniaps.neLdokNuload/maps/political-world-rnap.gif. Retrieved August 22, 2016.

Conclusions

Results presented in this study suggest the following:

1. Men and women are equally likely to be abducted and the age group ≤30 is more likely to be the age group of focus significantly more so than other age groups [(31-49) and (≥50)] examined:

    $$\chi^2 \, (2, N = 216) = 68.78, p = .000 \, (p < .0005).$$

2. The frequency of descriptions across six alien types observed during abductions was significantly greater than what would be expected by chance and the outcome favored the Humanoids as the species most influential in causing this overall outcome: $\chi^2 \, (5, N = 250) = 475.472, p = .000 \, (p < .0005)$.

3. The same relative pattern in Bullard's sequence of events was also found in this research indicating that abductions occur in a particular pattern (See, for example, Table 1, p. 32).

4. The following results suggest the possibility of literal events, but not conclusively so, for Hypothesis 2, Research Question 4: (a) Humanoids worked significantly more in the alone condition than in the together condition with other alien species when abducting humans: $\chi^2 \, (1, N = 170) = 9.412, p = .002$.

    (b) The North latitude, West longitude corridor (see Figure 3, p. 60) is the location significantly favored as the region of locations where results suggest alien beings choose to go on earth as they execute their intentions to abduct human subjects.

    (c) Total abduction scores in the range (48-67) significantly exceeded the frequency of such scores expected by chance compared to other score ranges. The scores in this range may be interpreted as a proxy for a series of unique actions experienced by abductees at the hands of their alien abductors in ways and for reasons known only to the abductors;

(d) The (3.1-7.0) range of derived GPS location values was the one geographical area of locations with a frequency found to be significantly greater than what would be expected by chance when compared to the frequency of other GPS ranges identifying locations where alien beings (i.e., the grays) likely visited/landed to abduct humans.

5. Statistical results comparing reported communication styles abductees experienced indicate a preferred mode of communication among aliens as they interact with humans during abduction events. That is, among the three communication styles reported by abductees in this study, *telepathy* and *no communication* were the styles of communication with frequencies greater than what would be expected by chance. The frequency for telepathy, however, was found to be significantly greater than the frequency for *no communication*: χ^2 (1, N = 175) = 7.000, p = .008. This result suggests that *Humanoid* types of aliens apparently favor the *telepathy mode* of communication.

6. The tripartite dilemma presented earlier in this paper (i.e., the technology of mankind, the technology of UFOs, and human abductions) is a relationship where two of its components are tentatively supported by the results of this study. First, while no information presented in this chapter can explain the nature or performance of UFO technology, it stands to reason, on the other hand, that if human witnesses to abductions (i.e., victims) are to be believed, as results on their psychological state from qualified professionals cited herein suggests, then abductees must have been aboard real physical objects . Moreover, abductees provided detailed descriptions of and responses to the internal environments of these craft (i.e., smells, colors, temperature, humidity, lighting, room/seating shapes or arrangements, etc.). Abductees also provided important information about the operators of these craft as indicated in items 1 and 2 listed below. Other findings in the list of items (1-9) are at the heart of the inextricable relationship stated previously in this research. That is, one cannot disentangle UFOs from abductions or vice versa, even if the operation and/

or propulsion capabilities of UFOs (or USOs, for that matter) cannot be presently explained.

7. The following additional items were characterized as highly likely outcomes due to results of statistical testing (i.e., p = .000 or p < .0005): The Abduction Total Score ranges (26-47) and (48-67). These ranges indicated the scores most representative of abductees.

8. Close encounters of the fourth kind involving 216 people in at least 30 countries and the statistical results of tested hypotheses and answers to research questions both indicate that close encounters of the fourth kind are highly likely to be literal events.

9. Standard Multiple Regression results (from data in Table 4) indicate that the regression equation with all three predictor variables (i.e., Work Condition, Region, and Score Range) was significantly related to G.P.S. location values as the dependent variable R = .687), R^2 = .471, adjusted R^2 = .462, F (3, 166) = 49.340, p = .000 (p < .0005). These results indicate the strong relationship between the range of *total abduction scores* (i.e., synonymous with alien eight-episode activity levels) *and geographical regions* favored in abductions world-wide.

Examining Descriptors of Abduction Events: Who They Are, What They Drive, Where They Go, and What Others Say They Do

CE-IVs maybe the key to the entire UFO mystery, but of all categories of reports, they are the most inherently unbelievable and the most difficult to verify. And to complicate the problem, the memories of witnesses to a CE-IV often seem to have suffered a strangely selective amnesia. In abduction cases particularly, most witnesses recollect only a close-up UFO sighting. A few may remember seeing alien creatures, but rarely do they recall many details. The actual contact or abduction experience has somehow been erased—perhaps mercifully so—from their conscious minds. Later, vague flashbacks, dreams, and intuitive feelings cause witnesses to suspect that something unusual has happened to them. And, nonetheless, details of the abduction experience remain locked in the deepest recesses of their minds. (Fowler, 2015, p. 27)

Who They Are

Chapter 2 explored four central questions: (1) Who are the abductors?; (2) What vehicles do they drive?; (3) Where do they choose to go on this planet as they carry out abductions?; and (4) What details are available about abductions as told by abductees? Who the likely abductors of humans are and how this information is known was addressed earlier, in part, by the alien descriptions on p. 26 of Chapter 1. The detailed information also stated how often six types of alien beings were seen. The test results from those frequencies indicated that the Humanoid species was by far the species most responsible for the Chi Square omnibus test results involving five other species (i.e., χ^2 (5, N =250) = 475.472, p < .0005). In terms of *who they are*, these and other results previously presented in Chapter 1which indicate the possibility of literal events in connection with close encounters of the fourth kind. Equally as important, the results suggest why abductees should be believed. Before moving on to the other three central questions mentioned above, let me address a second important factor related to the identity of abducting alien species. The omnibus result discussed above identifies Humanoids as the most observed species, but they are by no means the only ones. Table 10 presents a Percentile Rank (PR) analysis of scores, introduced in Chapter 1, shows the extent to which certain alien species tended either to work alone or work together. The alone and together work conditions were simply convenient ways to investigate the behavior of different alien species by comparing Total Abduction scores that were split into the working conditions shown in the above table. Incidentally, a gallery of alien descriptions by Huyghe (1996) and Story (2001) represent many of the descriptions and locations of accounts often reported by abductees. Thus, abductees' descriptions have informed the rest of us about their experiences as captured by the *eight episodes* discussed earlier by Bullard (1998). We should not forget, however, the negative impact on the bodies and minds of abductees as a result of their abduction experiences. Toward this end, the PR (Linn and Gronlund, 1995) approach employed here presents the transformation of Total Abduction scores (i.e., tallies of actions taken by alien captors) into a metric that represents how differently abductees say they were treated at the hands of their captors. In other words, the higher the Total Abduction score, the greater the degree of comprehensive actions defining an abduction experience. In turn, this also means that the higher the Total Abduction score, the more negative the experience had by abductees. The results of being captured were nearly always the same: many episodes filled with fear and often painful testing that served the agenda of the captors without regard for those who were captured. Moreover, these experiences

Table 10: Summary of Percentile Ranks (PRs) of (Total Abduction Scores) by Working Conditions

Alien Type*	Highest PR Alone	Lowest PR Alone	Highest PR Together	Lowest PR Together
Humanoids	99.55 (100)	2.12 (25)	99.63 (107) [Reptilians; and Nordic	4.53 (28) [Nordic; and Ordinary Humans] Reptilian/
Insectoid	57.85 (61)	21.01 (44)	99.63 (107) [Humanoids and Nordic]	11.63 (37) [Humanoids]
Nordic	-	-	99.63 (107) [Humanoids and Reptilians]	4.53 (28) [Humanoids and Ordinary Humans]
Ordinary Humans	2.45 (26)	2.45 (26)	5.65 (75) [Humanoids]	4.53 (28) [Humanoids and Nordic]
Hybrids	-	-	53.19 (60) [Humanoids]	10.59 (36) [Humanoids]
Monsters, Oddities, Robots & Apes	71.55 (67) [Humanoids]	19.45 (43) [Humanoids]	69.55 (66)	12.75 (38)

Note*: Names above from Bullard (see Clark, 1998).]

did not seem to get attenuated by changing the type of alien species conducting the invasive tests to which abductees were exposed. What was sometimes different, as seen in Table 10, was the extent to which abducting species of aliens worked either alone or together in pursuit of an agenda the exact nature of whichis still unknown. That being said, data in Table 10 indicates an overall relationship between humans and aliens (all of them) that was (and still is) *business as usual*. For example, readers will note that Humanoids worked slightly harder cooperatively with Reptilians and Nordics (i.e., Total abduction Score = 107) than when they worked alone (i.e., Total Abduction Score = 100). It is also clear that no other species *outworked* Humanoids when this species worked alone. One might conclude, therefore, that when an abduction occurred, an abductee most likely encountered Humanoids more often working alone than when they encountered other species. This might also support why grays are nearly uniformly the most familiar face reported (and what people think of) when the term *abduction* is mentioned. Table 10, however, extends this line up. Additionally, when most of the other five species did participate in abductions, they did so to a greater extent *cooperatively working with Humanoids* than when they abducted humans and worked by themselves. Only the last group (Monsters, Oddities, Robots, and Apes) worked harder alone than when they worked cooperatively with Humanoids. This might suggest a lower level of usefulness of humans to this collective group, but that is only a guess.

Nevertheless, Nordics were not reported as abductors of humans when working by themselves. Data in this study suggest that a variety of alien species abducted humans when working alone and did so as well when they worked cooperatively with Humanoids, Reptilians, Ordinary (looking) Humans, and Insectoid beings. Abductees often reported such interactions in various accounts analyzed for this study. Humanoids were either directly conducting examinations of humans or working cooperatively with other species to do so. Also, grays were seen in the background in a sort of supervisory capacity while other alien types conducted their own examinations of humans. The patterns emerging from an analysis of the data are clear: alien beings have an interest in humans and there appears to be both joint as well as individual interests. The psychological impact, however, remains unchanged: abductees have indicated that they were treated in the same manner that humans treat lab rats. Just like us, only the interests of alien beings, however, were important and not the concerns of those who were captured. Again, the analysis in Table 10 confirms what others have found previously but accomplished here using data derived from 216 abduction scores. Thus, the reliability and validity of Abduction Total Scores is self-evident.

Readers should note the number of countries listed below that are associated with claims of world-wide abductions. Were it not for abductees coming forward to tell their experiences, the rest of humanity possibly would not know the present extent to which the abduction phenomenon is occurring on a word-wide basis. If abductions were part of a murder scene, would not the bodies of victims be considered part of the evidence that a crime has been committed? Based upon what has been established and presented so far (1) *abductees are part of the physical evidence* so many insist is missing and (2) in addition to (1), it is highly likely that crimes have been committed against humanity (i.e., the abductions themselves). Given the documented trauma that abductees have experienced, their *minds, bodies, eyes, and ears* compositely shape a narrative that says to the rest of us, to a high degree of probability, (1) who the abductors are; (2) what abductors have most likely done; and, (3) are apparently, are still doing (i.e., multiple abductions of the same people)! Moreover, these victims sometimes have witnesses (i.e., the Linda Cortile Napolitano and Travis Walton cases, for example). More cases with physical evidence and, hopefully, witnesses, will be needed. Ideally, however, it would be better not to have any more abductions take place. Instead, have more cases, with witnesses, from the cases that have already occurred; if possible. It would be unethical to desire to have more abductions with or without witnesses and/or physical evidence just to get a greater number of ideal cases. A more humane treatment of abductees may likely increase the number of new and investigated cases which, in turn, could further legitimize abduction research. As shown below, the number of CE4 cases by gender indicates that (1) women were abducted in 12 countries across 89 cases and (2) men were abducted in 23 countries across 127 cases (i.e., 35 countries and 216 cases total):

Men

England–10; Argentina–4; Australia–8; Brazil–11; Austria–1; Chile–2; Canada–4; Israel–1; China–1; Spain–1; Sweden–1; Ireland–2; Morocco–1; Zimbabwe–1; Japan–1; Puerto Rico–1; Georgia (E. Europe)–1; Russia–1; Italy–1; USA–70; Germany–1; Poland–2; Mexico–1.

Women

England–2; Australia–19; Costa Rica–1; Germany–2; France–1; India–1; Canada–4; Poland–1; Finland–1; Sweden–1; Argentina–1; USA–55.

The 216 cases above only represent those included in this study out of at least 1200 or so cases that were in a variety of books, journal articles, and other sources collected for this research. Many of these accounts were credible sources but did not feature all of the variables needed for inclusion in the final 216 cases. Some of the criteria that eliminated many otherwise interesting cases included the lack of a date, city or location, time of day/night, and/or supporting investigatory information. Nevertheless, an examination of cases from the countries listed above offered an interesting picture of who these alien beings are. The exact nature of the end game among the various alien types, however, remains elusive. This missing component of the abduction phenomenon cannot be addressed without more cases that also increase the dearth of physical evidence. Nevertheless, a focused look at some of the evidence in hand, as being carried out here, will avoid the catastrophe conducted by the Battelle Memorial Institute. In 1955, NICAP (i.e., National Investigations Committee on Aerial Phenomena) and Colorado Project Member, David Saunders, reportedly expressed serious reservations about the way Battelle mishandled the requested statistical analysis in Project Blue Book Report Number14 (Dolan, 2002). According to Dolan, Saunders indicated that "with remarkable regularity, whoever did these statistics combined the categories so as to minimize his chances of finding anything significant" and, apart from the comments of Saunders, Dolan added, "to make matters worse, the original data, contained in IBM cards, had all been thrown away" (Dolan, 2002, p. 329).

What They Drive

The above information by country pertains to *CE4 cases* only. In this section of High Probability, **What They Drive** discusses 276 other cases involving UFOs (n = 210) and USOs (n = 66). The suspected relationship between shapes from CE4, UFO, and USO cases was examined for similarities in the descriptions of vehicles. This book is not the first to raise the possibility that all of these types of phenomena may be just one phenomenon. However, HP may be the among the few books that feature a primary emphasis on *calculated levels of probability* for the possible relationships between craft shapes and locations shared across reports of CE4, UFO and USO events. In all, a grand total of 458 cases were analyzed for possible relationships between CE4, UFO, and USO locations and craft shapes. The UFO and USO cases were collected from works by noted researchers and authors whose expertise and experience speaks for them (see, for example, pp. 73-74). To frame the analyses discussed later, a few descriptive details are provided here for each of the types of cases where *vehicle shapes* was the metric by which CE4, UFO, and USO cases were analyzed.

CE4 Cases

216 CE4 cases were reviewed to help answer questions about the types of vehicle shapes reported by abductees. As shown below, a total of 10 categories of shapes were reported 182 times across three frequency categories:

Shape Frequencies Seen

	Under 18 times	(19-38 times)		(39-59 times)	
Saturn shape	10	Light	35	Disc	56
Sphere	15	Domed Disc	21	Total = 56	
Cigar	10	Total = 56			
Cylinder	2				
Delta	6				
Other*	12				
Oval	15				
Total = 70					

*Note: Other includes—Diamond, Rectangle, Egg, & Football shapes.

UFO Cases

210 UFO cases were reviewed to determine how they might relate to close encounters of the fourth kind. The question here is whether or not objects seen in the same geographic locations share more than just proximity of observances. Or, is there a way to quantify how similar these similarities in location might be, since one group of objects we know of (i.e., as described by abductees) have been linked to human abductions to a high degree of probability. This question, of course, also will be asked about the extent to which geographic similarity in locations exists between CE4s and USOs, and between UFOs and USOs. Accounts of UFO sightings cover the range of witnesses that include everyday citizens to military personnel, airline pilots, police officers, air traffic controllers, and other persons not necessarily interested in promoting any particular narrative. Many cases also involved organizations and investigative journalists that have historically investigated these types of cases. Effect size (Cramer's V; (Pallant, 2010) was reported for all results. The authors whose works were analyzed for cases involving UFOs were (1) J. Allen Hynek (n = 58 cases), (2) Richard Dolan (n = 123 cases), and (3) Bruce Maccabee (n = 29 cases).

Shape Frequencies Seen

	Under 18 times		(19-38 times)			(39-59 times)	
Domed disc	12		Oval	25		Disc	81
Saturn shape	4		Light	33		Total =	81
Sphere	14		Cigar	20			
Cylinder	1		Total = 78				
Delta	4						
Other	16						
	Total = 51						

USO Cases

Sixty-six times, ten different USO shapes were observed in various geographical locations that have a proximity to the locations in which shapes of craft associated with CE4 cases also were observed. Is there a relationship between the shapes of unidentified submersible objects and the shapes of craft associated with their CE4 and UFO counterparts? And, if so, what is the nature of this relationship? Many of the reported USO cases involved military branches of governments, ordinary citizens at sea, commercial vessels, private research vessels, and persons in sensitive government positions. Sixty-six USO cases were reviewed from works by (1) Ivan T. Sanderson (n = 40 cases); (2) Paul Stonehill & Philip Mantle (n = 21 cases) and William Hamilton, III (n = 5 cases). References pertaining to the works of each of these authors are provided at the end of this chapter. A breakdown of the shapes of USOs observed coming out of and going into the waters of rivers, lakes, seas, and oceans of this planet is presented below:

Shape Frequencies Seen

	Under 18 times
Disc	14
Domed disc	2
Sphere	16
Cigar	10
Cylinder	3
Light	5
Oval	3
Other	13
Results	66

Observed Shape Frequency Totals

Experience	Under 18 times	(19-38 times)	(39 times and above)
(CE4)	70	56	56
(UFO)	51	78	81
(USO)	66		

It resulted in the following: χ^2 (4, N=458) = 119.888, p = .000 (p < .0005). A large Effect Size, Cramer's V = .362, was found to be highly significant (p = .000; p < .0005). This omnibus test result indicates that all three types of cases (i.e., close encounters of the fourth kind, UFO sightings, and USO sightings), share an overall significant relationship in terms of the frequency of craft shapes related to each case type. The overall magnitude of the relationship is highly meaningful, as measured by the effect size. Three follow-up tests were conducted to determine which case type experience was most responsible for the omnibus outcome. Note that the adjusted p-levels are less than .017 in each case (i.e., .05 ÷ 3 = .017). The three results obtained are shown below:

(1) **CE4 and UFO Shape Frequencies**: χ^2 (2, N =392) = 9.204, p = .010. Additionally, the Effect Size for this relationship was Cramer's V = .153, a small to medium effect size and significant result, p = .010. The outcome favored frequencies for UFO shapes as the source most responsible for the test result. CE4 and UFO case shapes were found to be significantly related to each other.

(2) **CE4 and USO Shape Frequencies**: χ^2 (2, N =248) = 74.063, p = .000; p < .0005). A large Effect Size, Cramer's V = .546, was found to be highly significant (p = .000; p < .0005). The outcome favored frequencies for CE4 shapes as the source most responsible for the test result. Across all three frequency ranges, the two types of case shapes were found to be significantly related to each other.

(3) **UFO and USO Shape Frequencies**: χ^2 (2, N=276) = 117.881, p = .000 (p < .0005). A large Effect Size, Cramer's V = .654, was also highly significant, p = .000; p < .0005). The outcome favored frequencies for UFO shapes as the source most responsible for the test result. UFO and USO case shapes were found to be significantly related to each other across all three frequency ranges. The above results were not totally unexpected, since

UFO *sightings* have been traditionally reported and investigated to a much greater degree than either USO or, particularly, CE4 events. What should not be overlooked, however, is the high probability associated with the relationship between CE4 shapes (i.e., described by abductees) and the shapes described by persons who witnessed either UFO or USO shapes under various circumstances world-wide. The relationships between the shapes of the three case types are at levels of probability that suggest the likely hood of literal events. Moreover, the shapes of these types of cases are related to each other with a high level of probability and pattern of consistency across shapes.

Where They Go

As most of us learned in school, the earth has four geographical regions: North latitude East longitude, North latitude West longitude, South latitude East longitude, and South latitude West longitude. There are two regions above and two below the equator. Hence, each time an event occurred, its location was defined by a G. P. S. location value developed for this study from traditional G. P. S. data. The resulting decimal numbers (i.e., X. Y Z) facilitated an easy way to correlate observed shapes. In this case, shapes were sorted using their derived G. P. S. data and the associated geographical region.

Geographical Regions

Shapes by Case Type	NlatElong	NlatWlong	SlatElong	SlatWlong
CE4	24	115	16	13
UFO	17	169	8	11
USO	20	27	5	14

The omnibus Chi Square Cross Tabs result above for the relationship between craft shapes by case types (using derived GPS location values for each shape) across four geographical regions was: χ^2 (6, N = 439) = 49.087, p = .000 (p < .0005). Cramer's V was found to be .236, indicating a medium to large effect size. That is, not only was the relationship between craft shapes by case types and geographical regions significant, but the strength of the relationship was moderate to strong (i.e., medium to large).

Additionally, each of the four variables were examined to determine if their separate shape frequencies were greater than the frequency expected by chance. Such information helped in understanding interpretations given to comparisons in other analyses where these variables may be correlated with each other. The Equal Frequency Hypothesis Tests applied to each of the four variables is shown below in Table 11.

The results of the four Chi Square analyses shown in Table 11 for (1) Case Type; (2) Observed Craft Shapes; (3) Geographical Region; and (4) G. P. S. Range of Values had frequency levels significantly beyond the frequency expected by chance. In each case, the probability of a chance occurrence *was not* the outcome supported by the results. The four outcomes favor a more literal interpretation in each instance ($p < .0005$). *Craft Shapes* were deemed a common thread connecting CE4, UFO, and USO cases. Color was also a variable briefly considered. However, because men tend to be more color blind than women, shapes were chosen instead.

Throughout this research, a concerted effort was made to examine details of each abduction case and make notes of instances where abductees might provide insights into two important cornerstones of the abduction phenomenon: (1) instances relating to physical evidence and (2) instances relating to psychological evidence of their abduction experiences. The views of several experts and/or experienced investigators have shaped important aspects of the debate about the reliability of abductees (i.e., see Hopkins, 1987; Jacobs, 1992;

Table 11: Equal Frequency Hypothesis Tests for Four Selected Variables

(1) **Case Type**	Observed N	Expected N	Residual
CE4	168	146.3	21.7
UFO	205	146.3	58.7
USO	66	146.3	-80.3
Total	439		

(2) **Observed Craft Shapes**	Observed N	Expected N	Residual
Disc	148	48.8	99.2
Domed Disc	34	48.8	-14.8
Sat. Shape	9	48.8	-39.8
Sphere	43	48.8	-5.8
Cigar	39	48.8	-9.8
Cylinder	6	48.8	-42.8
Delta	11	48.8	-37.8
Light	66	48.8	17.2
Other	83	48.8	34.2
Total	439		

(3) **Geographical Region**	Observed N	Expected N	Residual
NlatElong	61	109.8	-48.8
NlatWlong	311	109.8	201.3
SlatElong	29	109.8	-80.8
SlatWlong	38	109.8	-71.8
Total	439		

(4) **Range of GPS Location Values**	Observed N	Expected N	Residual
(0-3.0)	39	146.3	-107.3
(3.1-6.1) 330	146.3	183.7	
(6.2-9.5) 70	146.3	-76.3	
Total	439		

	Case Type	Craft Shapes	Geograph. Reg.	GPS Loc.
Chi Square	70.829	338.260	497.009	349.071
df	2	8	3	2
Asymp. Sig.	.000	.000	.000	.000

Mack, 1994; Bryan, 1995; Appelle, 1995/96, Fowler, 2015, and Laibow, 2016). The *other key witnesses* involved (i.e., the six groups on page 38) are not saying much. Without a resolution of the role of *all actors* in the abduction phenomenon, however, the truth being sought will be incomplete.

Nevertheless, those who do speak should be heard and attempts to understand should be made. In the following section, the voices of two abductees raise the possibility that they are telling the truth. To attack the messenger and *not examine* the message is a classic mistake. Apply Occam's Razor, then make up your own mind about who knows the most about this phenomenon: those who were captured or those who were not present. As the next section is read, consider the mystifying technology that was experienced and the shocking sight of approaching beings who were not human. A portion of the account from Hickson (H) and Parker (P) in the Pascagoula, MS Abduction Case makes clear the position of those who have had an experience they know happened, but also realize how difficult it will be to have others believe in their reality (Clark, 1998):

H: They better wake up and start believin' ...they better start believin'.

P: You see how that damn door came right up?

H: I don't know how it opened, son. I don't know.

P: It just laid up and just like that those son' bitches—just like they came out.

H: I know. You can't believe it. You can't make people believe it...

H: They won't believe it. They [are] gonna believe it one of these days. Might be too late. I knew all along they was people from other worlds up there. I knew all along. I never thought it would happen to me...(p. 716).

What Others Say They Do

The initial effort to look deeper into the experiences of abductees began by examining factors defined by the acronym PEPI (i.e., Physical Evidence (PE) and Psychological Impact (PI)). The PEPI factors in Tables 11 and 12 describe what abductees (1) *said they saw* and (2) what their *abductors did to them*. Additionally, in some cases, whether captors were seen or not, the captors were often heard (i.e., see forms of communication, p. 61, Chapter 1). Hence, in **What Others Say They Do,** physical and psychological dimensions of experiences had by abductees shape what is informative about the experiences of abductees. Many other parties such as investigators from ufological organizations, law enforcement, medical personnel, hypnotist(s), psychologists, and/or psychiatrists played a key role in representing what a given case presented. Nevertheless, the experiences of [101/127] men and [67/89] women were analyzed as much as possible according to their statements about their experiences. A repetitive number of acts experienced by abductees thereafter emerged. The final list of physical and psychological factors representing evidence of what happened are presented as PEPI total scores for each case. These scores were then analyzed to determine whichscores occurred at a frequency greater than the frequency expected by chance. Where needed, a secondary analysis was conducted to narrow the list of scores to a smaller group which might indicate the scores most representative of the overall experiences of abductees. The results shown in Table 12 indicate that PEPI scores and the other five variables exist at levels of probability that are most likely describing real events. The underlying pattern in PEPI scores, as well as in the other five variables shown, is the common physical and psychological experiences shared among male and female abductees. These people most likely saw the same craft shapes and types of beings no matter where they were world-wide. Readers are again reminded that whenever the probability level for a test is greater than $p = .05$, then the results are likely due to chance. Or, such a result means that the variables being analyzed are not very different from one another in terms of the frequencies describing them. A final look at PEPI factors in Chapter 3 will focus on the frequencies of items underlying the factors themselves.

Table 12: An Analysis of PEPI Scores of Males and Females in CE4 Events with Four Selected Factors

Gender

	Observed N	Expected N	Residual
Male	101	84.0	17
Female	87	84.0	-17
Total	168		

PEPI Scores

Score	Observed N	Expected N	Residual
15	6	16.8	-10.8
23	6	16.8	-10.8
31	16	16.8	-.8
39	**36**	**16.8**	**19.2**
46	**40**	**16.8**	**23.2**
54	**38**	**16.8**	**21.2**
62	16	16.8	-.8
69	6	16.8	-10.8
77	3	16.8	-13.8
85	1	16.8	-15.8
Total	168		

Observed Craft Shape

	Observed N	Expected N	Residual
Disc + Domed Disc + Sat. Shape + Sphere	95	56.0	39.0
Cigar + Cylinder + Delta +Light	46	56.0	-10.0
Other = (Diamond; Egg; + Rectangle + Oval + Football)	27	56.0	-29.0
Total	168		

Geographical Region

	Observed N	Expected N	Residual
NlatElong	24	42.0	-18.0

Geographical Region

	Observed N	Expected N	Residual
NlatWlong	115	42.0	73.0
SlatElong	17	42.0	-25.0
SlatWlong	12	42.0	-30.0
Total	168		

Observed Aliens

	Observed N	Expected N	Residual
Humanoids	99	56.0	43.0
Reptilians/Insectoid+ Nordic + Ordinary Humans + Hybrids + MORAS	20	56.0	-36.0
Independent Groups of Two, Three, or More	49	56.0	-7.0
Total	168		

Range of GPS Abduction Location Values

	Observed N	Expected N	Residual
(0-3.0)	15	42.0	-27.0
(3.1-6.1)	123	42.0	81.0
(6.2-9.2)	20	42.0	-22.0
(9.3-12.3)	10	42.0	-32.0
Total	168		

Test Statistics

	Gender	PEPI Scores	Observed Craft	Geographical Region	Observed Aliens	Range of GPS Values
Chi-Square	6.881	127.883	43.964	170.905	57.036	209.476
df	1	9	2	3	2	3
Asymp. Sig.	.009	.000	.000	.000	.000	.000

Conclusions

1. The Percentile Rank (PR) discussion presented in this chapter reveals which alien groups work alone and together with other alien species. Borrowing the conventional interpretation of the PR and applying it here simply means that if the PR for a particular alien species turned out to be 90, this calculated value would refer to the percent of other alien groups that scored lower than the species with a PR of 90. Similarly, only 10 percent of other groups would have a possible PR higher than a group with a PR of 90. Thus, from one score we get a sense of the effort some alien species are putting into their abduction efforts relative to other groups by comparing their PR values. The results of PR calculations for all species identified by abductees are shown in Table 10. Also discerned from the table is which species worked alone or together with the highest PR (and lowest PR) in an effort to accomplish their unknown agenda. The PR of the Humanoid species lead in both the Alone and Together work conditions compared to other species in this study using abduction score totals derived from abduction accounts.

2. The results of analyzing craft shapes and geographical location relationships among CE4, UFO, and USO cases reveals a highly significant relationship. This implies that the craft shapes involved in CE4 cases and the craft shapes in both UFO and USO cases and their respective locations overlap as a basis of their relationship. If UFO and USO craft are not involved in abductions, why does data for their locations and shapes correlate so highly with craft shapes and locations related to CE4 cases? The N latitude, West longitude region of the planet was a common geographical region for CE4, UFO, and USO cases: χ^2 (6, N = 439) = 49.087, p = .000 (p < .0005). These results indicate that the odds of this outcome being due to chance is less than 1 in 1,000.

3. Various *craft shape totals* and how often they were observed across three frequency categories were analyzed using the Cross Tabs procedure for Chi Square. This produced highly significant omnibus results: χ^2 (4, N = 458) = 119.888, p = .000 (p < .0005). Each of the follow-up tests also produced significant results: (1) χ^2 (2, N = 392) = 9.204, p = .010 for *CE4 and UFO Shape Frequencies*; (2) χ^2 (2, N = 248) = 74.063, p = .000 (p < .0005) *for*

CE4 and USO Shape Frequencies; (3) χ^2 (2, N = 276) = 117.881, p = .000 (p < .0005) for *UFO and USO Shape Frequencies*. Disc and Domed Disc craft shapes made the largest contribution to the above omnibus test results. Thus, the shapes of these types of cases are related to each other with a high level of probability and pattern of consistency.

4. With respect to specific G.P.S. location ranges and craft shapes for all case types, the (3.1-6.1) range of GPS location values was found to be the range of G. P. S. location values that most contributed to the omnibus outcome reported here: χ^2 (2, N = 439) = 349.071, p = .000 (p < .0005). The (3.1-6.1) G.P.S. Range of Values includes locations (31-45)° N Latitude and (0-15)° W Longitude to (76-90)° N Latitude and (0-15)° W Longitude. This includes most of the population centers 30° above the equator to the North Pole world-wide. This means there is a high probability that a significant slice of earth's population is being visited world-wide.

CHAPTER 3

Consistency of the Narrative

Ｉt is only in the last few years of human genetic research that skin sampling proce-dures have become common as an efficient way of collecting useful genetic material. Is it possible that the decades-old scoop marks and skin scrapings were signs of an advanced alien technology preceding our own, in the way that Betty Hill's "pregnancy test" was a precursor of amniocentesis? (Hopkins & Rainey, 2003, p. 147).

The experiences of abductees tend to be described by a redundancy in related and specific acts that are physical and psychological. Some acts are active, and some are passive on the part of both abductees and their captors. A most disturbing element accompanying many abduction accounts is that in spite of moments of resistance by abductees, the vast majority of these accounts clearly indicate that the abductors were in full control. Not even being armed made a difference at any point during an abduction. These beings have the capacity to do whatever they wish in pursuit of their goals and then they leave. If they wish, they come back again and again depending upon whom it is that is the subject of their interest. Exactly why this occurs is not known. Abductees who have experienced being abducted once or more than once report such interactions.

Nevertheless, the 1024 statements presented in Table 13 represent how often 167 abductees experienced the physical and psychological acts shown. Some were left with scoop marks while others say they had human sperm and eggs removed. Still others had images of their captors sealed in their memories which they found impossible to forget. These images were later drawn in sketches because the abductees were wide awake! Others who endured the same sequence of abduction episodes agreed to be hypnotized but revealed sketches with the very same features of their captors. The number of physical and psychological actions identified here provide a collective sense of what abductees endured. Hence, it is important to know if the frequency of what abductees described occurred at a level greater than the frequency expected by chance. If these things occurred by chance, so be it. But if not, the next step is to look closer at even more cases to see if a pattern can be established. The consistent occurrence of these acts indicates that something extraordinary is happening on a world-wide scale. The statistical determination of the level of *likely hood* associated with what abductees have reported is equally extraordinary. Think about that!

The frequency of descriptions of physical and psychological actions reported by abductees was significantly greater than what would be expected by chance: $\chi^2(9, N = 1024) = 192.523$, p = .000 (p < .0005). Note the descriptions shown below that collectively were the most influential in contributing to the omnibus outcome. The odds that these events occurred by chance are 5 in 10,000 or about 1 chance in 1,000. The results shown in Table 13 favor the likely hood of real events having occurred in the categories of actions shown.

The short list of PEPI items that most contributed to the results in Table 13 are shown below. While Item 5 below appears frequently on book covers and prominently in sketches by abductees, note as well the even *larger frequency* of high strangeness items *1, 6, and 7* that have been compared to the more familiar Item 5.

Item Number/Physical Actions	Item Number/Psychological Actions
1= altered human female; fetus missing;	5=Can't forget eyes/body of hairs from alien-human sexual act beings
3= Abducted 1x or more	8=Fear
6=Missing time; watch not working; incorrect time	10=Dreams; flashbacks; hypnotism sought/used
7=Abductees float; light elevates body to ship; moved thru air/walls/closed windows	

The three testimonies following Table 13 are insights offered to reveal the complex range of experiences that are branches of the same abduction tree. That is, the testimonies offer a broad context of what abductees experienced. These accounts also reflect the psychological and physical items in Table 13 and the broader themes in Table 14. How is it that people all over the world are saying the *same universe of things* about a type of experience that is not supposed to be real? The findings presented so far are telling the king he does not have on any clothes.

Table 13: Physical and Psychological Subcategories of Acts Describing Abduction Events

	Descriptive Statistics			
N	Mean	Std. Deviation	Minimum	Maximum
1024	5.62	2.783	1.00	10.00

Subcategories (P = Physical; Psy = Psychological)

		Observed	Expected	
Item	Description	N	N	Residual
1-P	Altered human female; fetus missing; punch biopsy; implants; objects left in body; hair from human-alien sexual encounter	111	102.4	8.6
2-Psy	Faints; collapses	31	102.4	-71.4
3-P	Abducted 1x or more	167	102.4	64.6
4-P	Sketches craft/beings	53	102.4	49.4
5-Psy	Can't forget eyes/body of beings	110	102.4	7.6
6-P	Missing time; watch not working; incorrect time	132	102.4	29.6
7-P	Abductee floats; light elevates body to ship; moved thru air/walls/closed windows	117	102.4	14.6
8-Psy	Fear	145	102.4	42.6
9-Psy	Connects w/aliens; develops sense of *mission*	35	102.4	-67.4
10-Psy	Dreams; flashbacks; hypnotism sought/used	123	102.4	20.6
	Total	**1024**		

Test Statistics	
	PEPI Subcategories
Chi Square	192.523
df	9
Asymp. Sig.	.000 (p < .0005)

Testimonials as Insights into Abductions

(1) We (abductees) are caught in the middle of a series of ongoing events that will one day prove to be historically significant to the future of the earth. Unfortunately, because of our continuing, non-voluntary interactions with these beings, we have, in a sense, become victims of both sides of the phenomena. On the one hand, we are used by aliens for their own purposes, whatever those purposes might be. On the other, if we speak out, we risk ridicule from the public, and in some cases, possible harassment and/or intimidation by government and military personnel. (Linda Porter, 1991, pp. 1-2 of 9; February 28 postmarked letter to Linda Moulton Howe).

(2) I'm standing up on nothing. And they take me out all the way up, way above the building. Ooh. I hope I don't fall. The UFO opens up almost like a clam and then I'm inside, said 41-year-old (at the time) Linda. I see benches similar to regular benches. And they're bringing me down a hallway. Doors open like sliding doors. Inside are all these lights and buttons and a big long table. I don't want to get up on that table. They get me on the table anyway. They start saying things to me and I'm yelling. I can still yell. One of them says something like [Nobbyegg]. I think they were trying to tell me to be quiet because he put his hand over my mouth. (Linda Cortile Napolitano Case; Hopkins, 1992, p.1 of 4).

(3) Ed: I sense fear and I look around and I'm, like, floating. I don't see Doug, I don't see Doris. This is happening so fast. [A long pause] Now I can see. I'm looking down and I see they're in the backseat of the car and it's like they're asleep. I'm sort of going up, further, further away.

BH: Okay, let's go back to when you feel that you are moving away from them, from Doris and Doug. You can see them through the car window], but let's go back to your feeling of movement. How were you moving? Down the street or moving back towards the Outback? Which direction are you moving in?

Ed: Up. It [the car] is getting further away and I see something.

BH: What do you see?

Ed: I see something metallic and it's shaped like a ... looks like a shovel. Well, the head is shaped like...no, it's more like a ...it's hard to describe. Like a giant arrow.

BH: Sort of pointed more on one end?

Ed: Yes, but the shaft is very thick. [Ed responds to other questions from Hopkins; brackets my own].

Ed: It's like I'm lying down...I'm lying on a ...it seems like lying on my back. But I don't feel anything. My thoughts are strange.

BH: What are your thoughts?

Ed: That...why is this happening to me again? (Hopkins & Rainey, 2003, pp.269-270).

Insights to Consider

The first quote clearly communicates how many abductees feel about the life circumstances they must endure without much help or sympathy from people they feel should be taking them seriously or acting in a more supportive manner. This also may be, in large part, why many abductees choose not to make their accounts known or agree to have them thoroughly investigated beyond initial disclosure. They understand what happens to some people who do.

The second quote spotlights a number of things characterized in the literature as *high strangeness*. Note the reference to *standing on nothing* as Linda Cortile Napolitano indicates that she was *floated* out of her apartment some 12 stories [i.e., through a closed window in the full version of her experience; brackets my own] (120 feet) above the ground in her nightgown. She is awake! Moreover, her abduction was witnessed by four people!! One was Javier Perez de Cuellar, former Secretary General to the United Nations. Others included two of his bodyguards who brought the matter to his attention. The fourth person was a woman named

Janet Kimble (Kimbell) who was on the Brooklyn bridge and also witnessed Napolitano floating through the air and into a waiting craft (Hopkins, 1992).

The third quote demonstrates a couple of key points about how a careful hypnosis session is conducted. BH is Budd Hopkins, if you didn't already know. One obvious strength in the Hopkins approach is the presentation of the transcript revealing how the session proceeded and what the client said and felt in his/her own words in response to questions from the hypnotist. A second point is that there were no leading questions used during the sessions. That is, note the way that Ed describes the direction he is moving relative to the way Hopkins asks about his movements. In the twenty-one or so references cited in this book, Hopkins maintains this same manner of questioning. This noted author does not need my endorsement of his methodology. I make these comments because my use of his work required that I see evidence of how consistently he lived up to his own enunciated standards for conducting hypnosis sessions. I then read and reasoned on my own. This assessment also helped to educate me regarding how hypnosis sessions by Hopkins compared to those conducted by other investigators in other abduction accounts. Critics of hypnosis advocate that all hypnotists operate to lead witnesses toward some preconceived outcome and, thereby, find what the hypnotist is looking for in such sessions. Clearly, this is not the case. A more interesting set of questions to pursue might be: (1) What kind of advanced technology *floats* a person through the air; or through walls and closed windows?; (2) Who or what is operating the craft aboard which people were transported as described in various cases?; (3) What agenda is being pursued? Some of these questions are partly addressed by the information presented earlier in Chapter 1. But clearly all issues are not settled. On the other hand, the fantastic levels of technology, if ever understood by our science, will undermine claims by those who scoff at such technology belonging to alien beings, since such critics claim that such technology is probably black operations rooted; a dubious claim at best in light of there being no physical evidence or tested theory presented by any (humans) anywhere in the world (i.e., remember that requirement?). At any rate, if any of the space faring nations on this planet or nations planning to be space faring had such technology, it is my prediction, based upon how rapidly nations expose their advantages, that such a nation (or nations) would have shown the U. S. and the rest of the world that they have conquered gravity and, perhaps, inertia as well. Demonstrating that one has made such a quantum leap in science and technology is the true test of being able to *float individuals* twelve stories above ground, silently, into a hovering ship while several witnesses watched (i.e., the Linda Cortile Case, for example). Mankind is simply not there yet. If we were, mankind would not rely on chemically propelled rockets as we currently do world-wide. If you hold a different opinion, fact check me; show me the way. My email address is at the end of this book.

In this study, 1024 items describing physical and psychological forms of evidence came from 67 out of 89 abduction cases that involved women and 100 out of 127 cases that involved men. The high probability of the statistical testing presented below underscores the literal nature of the experiences of abductees world-wide. The overall analysis in this chapter reveals key insights into the consistency of close encounters of the fourth kind and further highlights what it is that *abductees say these beings do* during abduction events. In addition to the probabilities presented, abductees' minds and bodies, as described in the frequency of statements related to their minds and bodies, show a level of consistency in the descriptions provided by abductees.

Table 14 and the Chi Square omnibus results indicate the broad-based themes most responsible for the highly significant outcome found: χ^2 (N = 489, 12) = 241.260, p = .000 (p < .0005). Several of the themes had Observed Ns less than their Expected Ns and did not contribute to the eventual result. The broad *physically oriented themes* where Observed N's were larger than Expected N's were items **4, 8, 10 and 12.** The *psychologically oriented themes* where the Observed N was larger than the Expected N was item **7.** Clearly the most dominant theme was the emphasis on reproduction among aliens whether working alone or in cooperative groups during examinations of humans. Various accounts read for this research involved arranged sexual episodes involving humans and aliens aboard ship or the presentation of hybrid off-spring allegedly the product of human eggs taken from women during some previous examination and, thereafter, later fertilized by alien means to produce hybrid off-spring. These variety of reproductive concerns (i.e., some even involved showing human women hybrid babies and inferring that they were the mothers) were acts revealed by human males and females. Sometimes these events involved the assistance of hypnotists to help men and women learn the extent of the abduction experience. It is also obvious, that not all the men and women had to be hypnotized to tell what they experienced (i.e., awake men + women = 71, while men + women hypnotized = 81). It is also an interesting comparison to note that *fewer men than women* experienced their abductions while awake (i.e., 23 compared to 48). Perhaps men had to be subdued because they posed a threat of some greater degree than women to their captors or because men reacted more violently than women did. This seems to be plausible, since many accounts of abductions, the Betty and Barney Hill case, for example, details how Betty walked aboard the ship escorted by aliens, while Barney was assisted aboard and off the ship with alien beings holding him up as though he was *sleep walking*, according to Betty Hill (Fuller, 1966, pp.159; 177). This was also true among some members of the five Allagash abductees. For example, Jim Weiner's Chapter 3 description of how the men were beamed aboard ship through a portal of light, examined, and moved around inside the craft with assistance while under the *direction and control* of alien captors. Such descriptions were similar among the five men. Hypnosis helped to expose their similar stories (Fowler, 2005). Chuck Rak's Chapter 6 description of being *placed back in their canoe* by alien beings (Fowler, 2005, pp. 142-147) was also consistent with other accounts in this study. The above numerical comparisons of how men and women were treated raises questions as to why *fewer women than men* (i.e., 28 compared to 53) had to be hypnotized to unveil experiences they say were hidden from them by their captors. Why this happened is, of course, unknown. But the element of control of abductees and the control of details of the abduction itself by alien beings are details that have been revealed by abductees. In many cases, abductees were awake and in other cases abductees who were hypnotized revealed the same controlling actions by their alien captors. Either way, the consistency of the narrative is remarkable and many of the broad themes of the abduction narrative are supported by a highly significant probability which indicates that the overall outcome was not due to chance.

Table 14: Broad Physical (P) and Psychological (Psy) Themes Characterizing Abductions

Number and Theme	Observed N	Expected N	Residual
1 = Save the planet (P	13	37.6	-24.6
2 = Save human beings (P)	6	37.6	-31.6
3 =Enlightenment; love (Psy)	5	37.6	-32.6
4 = Reproductive concerns; alien women made pregnant by humans; human women impregnated by aliens (P)	103	37.6	65.4
5 = No idea about abductions (Psy)	27	37.6	-10.6
6 = Men awake; 1 alien specieson board ship (P)	23	37.6	-14.6
7 = Men hypnotized to understand abduction (Psy)	53	37.6	15.4
8 = Women awake; 1 alien species on board ship (P)	48	37.6	10.4
9 = Women hypnotized to understand abduction (Psy)	28	37.6	-9.6
10 = Men alone on board with 1 alien species (P)	64	37.6	26.4
11 = Men on board with more than 1 alien species (P)	37	37.6	-.6
12 = Women alone aboard ship with 1 alien species (P)	59	37.6	21.4
13 = Women together more than 1 alien species (P)	23	37.6	-14.6

Total = 489

Conclusions

1. 1024 statements from 167 abductees were analyzed using the Chi Square Equal Frequency Hypothesis. The results in Table 13 indicated that seven of the ten factors made the largest contribution to the omnibus test results found: χ^2 (9, N= 1024) = 192.523, p = .000 (p < .0005). Four of the factors were physical acts: (1) Altered human female; fetus missing; (3) Abducted 1x

or more; (6) Missing time; watch not working; and (7) Abductee floats; light beam elevates body to ship; moved thru air/walls/closed windows. Three of the factors were psychological in nature: (5) Can't forget eyes/body of beings; (8) Fear; and (10) Dreams; flashbacks; hypnotism sought/used. The odds that these seven factors occurred by chance were calculated to be 1 chance in 1,000. These results favor the likelihood of real events being described by the actions involving abductees and their captors.

2. Three quotes from abductees (Porter (letter to Howe),1991; Hopkins, 1992; and Hopkins & Rainey, 2003) demonstrated the psychological trauma of living as an abductee and two other quotes focused on the high strangeness so typical of abductions where persons are *floated* up to a waiting craft. One of these abductions (i.e., the Cortile-Napolitano case) was witnessed by *four observers*. This fact destroys arguments against the existence of technology that has so far not been demonstrated to be a product of human engineering or back engineering. Since this event happened to one person with witnesses, why can't it happen to one person who has no witnesses? Logic is not on vacation here. What the abduction phenomenon illustrates is that more scientific effort is needed to provide a greater degree of understanding of this world-wide phenomenon. Chance is clearly not always the favored explanation, as analyses in this chapter show. Those who actually experienced the phenomenon are the sources of data under examination in this book. Toward this end, Chapter 5 will explore more closely why abductees should be believed.

3. Moreover, an analysis of *broad themes found across the 216 experiences* of ordinary people whose abductions were their common links indicates that a repetitive pattern exists: (a) Reproductive concerns of alien beings involved producing hybrid children from eggs taken from human women who see these children during an abduction event and are forced to hold/provide nurturing behavior as aliens looked on; (b) Men, more so than women, employed hypnotism to help understand their abduction experience; and (c) Women, more so than men, were awake during the abduction experience with at least one alien species onboard a craft.

4. At nearly the same frequency, men and women observed that only one alien species was on board a craft during their abduction.

5. Previous analyses have established that the Humanoid species (i.e., grays) was the species most often involved in abductions.

6. The level of probability dominating results of Chi Square tests indicates that nearly all outcomes are occurring at levels far below the .05 level of probability. In fact, the levels are consistently found to be about 1 chance in 1000 that chance was the best explanation for the results obtained.

An Analysis of Relationships Among G.P.S. Location Ranges, Alien Groupings, and Craft Shapes

In this research, what abductees say happened to them, especially when they were awake, is important since they are the eyewitnesses present with their abductors. Not all abductors, however, looked the same. Provided below are two descriptions given by an abductee, Travis Walton:

(1) Travis saw three figures dressed in loose-fitting orange one-piece suits standing near him...As his vision cleared, he recoiled in shock and horror as he realized these were not human beings. He would describe them this way: They were short, shorter than five feet, and they had very large, bald heads, no hair. Their heads were domed, very large. They looked like fetuses. They had no eyebrows, no eyelashes. They had very large eyes—enormous eyes— almost all brown, without much white in them. The creepiest thing about them were [sic] those eyes. Oh, man, those eyes, they just stared through me. Their mouths and ears and noses seemed real small, maybe just because their eyes were so huge. [Barry, 1978]. (Walton Abduction Case, Clark, 1998, p. 993).

(2) The man looked like a deeply tanned, muscular Caucasian, about six feet two inches tall, perhaps 200 pounds. He had sandy blond hair long enough to

cover his ears, and he was dressed in a tight-fitting, bright blue coverall suit with a Black band or belt across the middle. He wore black boots. In his excitement Travis failed to appreciate just how odd the man's eyes looked. A "strange bright golden hazel," they were not really the eyes of a human being. (Walton Abduction Case, Clark, 1998, p. 994).

Table 15: G.P.S. Ranges of Value and Their Relationship to Three Alien Groupings

Code and G.P.S. Range	Frequency by G.P.S.	Range Group
1 = (2.0-4.0)	65-------1s	(1) Humanoid
	9-------1s	(2) Other Aliens (no Humanoids)
	35-------1s	(3) Groups of Two or more
	Total =109-------1s	
2 = (4.1-6.1)	22-------2s	(1) Humanoids
	1--------2	(2) Other Aliens (no Humanoids)
	7--------2s	(3) Groups of Two or more
	Total = 30--------2s	
3 = (6.2-8.2)	5--------3s	(1) Humanoids
	2--------3s	(2) Other Aliens (no Humanoids)
	3--------3s	(3) Groups of Two or more
	Total = 10--------3s	
4 = (8.3-15.0)	9----------4s	(1) Humanoids
	7-------4s (2)	Other Aliens (no Humanoids)
	3--------4s (3)	Groups of Two or more
	Total = 19-------4s	

Grand Total = 168

In this chapter, the initial objective was to use derived G.P.S. location values as a way to sort all abduction cases involving individual aliens or groups of aliens reported as abductors. That is, the G. P.S. values were sorted across four categories [i.e., (2.0-4.0), (4.1-6.1), (.6.2-8.2), and (8.3-15.0)]. Since the alien abductors in these cases were already tied to G.P.S. case data, this procedure effectively sorted the following three categories of alien beings across the four ranges above. The abducting alien groups were (1) Humanoids; (2) Other Alien Types, excluding Humanoids; and (3) Groups of Two or more, including Humanoids. Frequencies associated with each of the three groups formed a data set that was then analyzed using the Chi Square Equal Frequency Hypothesis. A breakdown of how many times each alien group was observed is shown in Table 15 above.

The Equal Frequency Hypothesis Test results indicated that the highest frequency involved the Humanoid species in the G. P. S. range (2.0-4.0) or (16-30)°N Latitude and (0)°W Longitude to (46-60)° N Latitude and (0)°W Longitude from the Prime Meridian. The calculated Chi Square value was: χ^2 (3, N = 168) = 147.286, p = .000; p < .0005. The probability that these results were due to chance is less than 1 in 1,000. Hence, the results favor an outcome which says that it is highly likely that abductees saw the alien beings that abducted them in the location ranges involved. Rearranged data from Table 15 for all ranges is shown below:

Range	Observed N	Expected N	Residual
1 = (2.0-4.0)	109	42.0	67
2 = (4.1-6.1)	30	42.0	-12
3 = (6.2-8.2)	10	42.0	-32
4 = (8.2-15.0)	19	42.0	-23
Total = 168			

By omitting the first range and recalculating Chi square, it was found that G.P.S. values in the (4.1-6.1) range also involved *Humanoids as the species* most responsible for the omnibus outcome: χ^2 (2, N = 59) = 10.203, p = .006. The Chi Square result favoring the Humanoid alien group (only) versus the other two alien groupings in the **(4.1-6.1)** range of values was found to be: χ^2 (2, N = 59) = 10.263, p = .006. Results for the remaining two G.P. S. Ranges of Location Values and the three alien groupings was: χ^2 (1, N = 29) = 2.793, p = .095. Thus, alien beings seen abducting humans across the G.P.S. Ranges [(6.2-8.2) and (8.3-15.0)] were equally likely to be involved in abducting humans in these locations.

The Relationship Between Craft Shapes and G.P.S. Locations. The shapes of craft across CE4, UFO, and USO case types and the G.P.S. locations of these shapes were examined in order to address the following conundrum: If craft shapes with grays were found in particular G.P.S. locations at a certain frequency, what accounts for craft shapes identified as UFOs or USOs being present in the same ranges of GPS Value locations? It seems reasonable that in none of the cases in this study were alien beings seen arriving or leaving by means of walking to get from one location to another. Craft moving from one location to another and returning were reported by abductees. They were also reported by persons who only observed and reported either UFOs or USOs. Therefore, craft shapes for these three case types and the locations where they were reported is a viable relationship to explore. The frequencies of these shapes *by case type* and the Chi Square results for an Equal Frequency Hypothesis test are shown below:

G.P.S. Ranges of CE4 Craft Shapes

	Observed N	Expected N	Residual
(2.0-4.0)	96	33.5	62.5
(4.1-6.1)	14	33.5	-19.5
(6.2-8.2)	8	33.5	-25.5
(8.3-15)	16	33.5	-17.5

Total = 134

Test Statistics

Chi Square = 156.507   df = 3   Asymp. Sig. = .000 (p < .0005)

G.P.S. Ranges of UFO Craft Shapes

(2.0-4.0)	151	53.0	98.0
(4.1-6.1)	42	53.0	-11.0
(6.2-8.2)	8	53.0	-45.0
(8.3-15)	11	53.0	-42.0

Total = 212

Test Statistics

Chi Square = 254.981   df =3   Asymp. Sig.= .000 (p < .0005)

G. P. S. Ranges of USO Craft Shapes

(2.0-4.0)	35	16.5	18.5
(4.1-6.1)	11	16.5	-5.5
(6.2-8.2)	9	16.5	-7.5
(8.3-15.0)	11	16.5	-5.5

Total = 66

Test Statistics

Chi Square = 27.818   df = 3   Asymp. Sig. = .000 (p < .0005)

They All Look the Same

The analysis presented next emphasizes an important characteristic about CE4, UFO, and USO craft shapes: each of the three categories of events *have craft that all look the same.* Except for those rare instances where a witness or witnesses observed a person being taken aboard a waiting craft, the only other distinguishing feature separating CE4 cases from the other types of cases has been the eyewitness accounts of abductees themselves. Otherwise, CE4 craft could just as easily be called a UFO or be labeled a USO. An object or craft described in a CE4 event could be labeled a UFO if seen prior to or following an abduction. If the same object or craft were observed just before or after an abduction going into or coming out of a body of water, it could justifiably be labeled a USO. Given these circumstances and the apparent commonality of reported similar shapes across the above categories, it was important to determine if any relationship could be determined based upon (1) a commonality of shapes *without regard* to the source of the shapes and (2) a commonality in shapes that involved the sources of the shapes as well as common locations where the shapes were reported. Location of shapes reported is the other common thread for all types of events. The first analysis presents what relationships were found with respect to shapes and locations *regardless of case type.* These correlations suggest that the sources of shapes is not enough to explain why these shapes regardless of source are in such proximity to each other. It is highly likely given the above results that shapes belonging to UFO or USO events could be the same shapes associated with CE4 cases. Perhaps, the underlying explanation for these highly significant correlations is that the unknown purpose of the various craft shapes is what separates them and not the labels assigned by man. The purpose of these similarly shaped craft seems to be a key missing element that, if ever known, could answer a lot of questions regarding why these shapes have been reported in the same locations, in spite of the distinctions defined by their labels. As shown above, the correlations do not favor coincidence or chance as the preferred explanation.

The second important aspect of the relationship between categories of shapes, frequencies of observed shapes within categories, *and the locations* where the shapes were reported is shown next.

G.P.S. Value Ranges by Case Type and Craft Shapes. An analysis was conducted that focused on sorting derived G.P.S. data of craft described in UFO and USO cases across four G.P.S. Ranges of Values locations. The G. P. S. location data and the same four ranges also were used with data involving CE4 cases. This analysis sought to determine if *location data* for UFO and USO craft shape frequencies share a relationship with the *location data* of craft shape frequencies associated with CE4 cases.

The Equal Frequency Hypothesis approach established that (1) the frequency of shapes for **CE4 cases** was highest in the range (2.0-4.0) with a calculated value of χ^2 (3, N = 134) = 156.507, p = .000; p <.0005); (2) the frequency of shapes for **UFO cases** was highest in the range (2.0-4.0) with a calculated value of χ^2 (3, N = 212) = 254.981, p = .000; p < .0005); and (3) the frequency of shapes for **USO cases** was highest, as well, in the range (2.0-4.0) with a calculated value of (χ^2 (3, N = 66) = 27.818, p = .000; p < .0005). This pattern occurred in three out of the four G.P.S. ranges with shapes in those locations across all three case types. As can be seen above, the results for each case type were not due to chance (i.e., odds favoring chance were calculated to be 1 in 1,000). Perhaps the beings observed aboard CE4 craft have some common interest with whomever or whatever is controlling UFO and USO craft. Whatever this common interest is, beings aboard CE4 craft seem to share this interest in a common set of locations world-wide with both UFO and USO craft. It may not be unreasonable, therefore, to assume that one possible explanation for this proximity of presence is that the similarly shaped craft in the same locations over time may be an indication that all such craft are engaged in the same activity...human abductions. This latter assumption also implies that just as similarly shaped CE4 craft and their alien pilots have been described by abductees, perhaps the similarly shaped craft in UFO and USO cases in these same locations are being piloted by the same assortment of beings observed aboard CE4 craft. Thus, craft shapes of UFO, USO, and CE4 events may be erroneously understood as separate entities when, in fact, the highly significant statistical results here suggest that the craft shapes from three different sources are *highly likely the same group of craft*. I wonder why there would be two groups of craft that simply fly around planet earth and not worry, apparently, about being seen, but at the same time, be unconnected to abductions. And, yet, a similarly shaped group of craft that fly about planet earth *are described by abductees* as craft shapes aboard which they saw various beings that were not human. One connection is supported by the above correlations and that is that *all of these shapes share common locations world-wide* where humans have been abducted at levels of probability that *do not favor chance* as the best explanation for why these craft shapes share such a significant relationship with locations, case types notwithstanding.

Table 16: Bivariate Correlation of Three Groups of Craft Shapes by Four G.P.S. Range of Values

G.P.S. Value Ranges	Disc; Domed Disc; Saturn Shape; Sphere	Cigar; Cylinder; Delta; Light	Other Shapes: Rectangle Egg; Oval; Football
(2.0-4.0)	52	30	14
(4.1-6.1)	9	5	1
(6.2-8.2)	6	2	
(8.3-15.0)	11	4	

Results

Shapes	Correlation	Result (r =)	Sig. (p=)	Bonferroni Correction
Disc; Domed Disc; Saturn Shape; Sphere	**w/**Cigar; Cylinder Delta; Light	.998	.002	.016
Disc; Domed Disc; Saturn Shape; Sphere	**w/**Other Shapes: Diamond; Rectangle; Egg; Oval; Football	1.000	.01	.016
Cigar; Cylinder; Delta; Light	**w/**Other Shapes: Diamond; Rectangle; Egg; Oval; Football	1.000	.01	.016

Table 17 presents the results of Chi Square tests that sought to examine the nature of the relationship between CE4, UFO, and USO G.P.S. shape locations and how they might be correlated across four ranges of G.P.S. ranges. The omnibus results of the above Chi Square tests strongly suggest that CE4, UFO, and USO craft shape frequencies *share a highly significant relationship* with the four world-wide Ranges of G.P.S. Values. This essentially means that across various locations on this planet, particularly with respect to where abductions have occurred (i.e., CE4 cases), the G.P.S. locations of vehicles associated with UFO sightings and USO sightings share a significant relationship with the G. P. S. locations of craft shapes associated with CE4 cases. Another suggestion is that *UFO and USO craft may be linked to abduction activity* based upon the highly significant relationship between the G. P. S. locations of UFO and USO craft shapes relative to the G. P. S. locations of craft in CE4 cases. To pinpoint which case types are mostly responsible for the omnibus results in Table 17, it was necessary to perform secondary or follow-up tests shown in the **Secondary Chi Square Testing** section of Table 17. Secondary test results clearly show that *UFO craft shape frequencies* across four ranges of G. P. S. locations were mostly responsible for the test results for *all case types*. UFO and CE4 and UFO and USO shape frequencies across four G. P. S. Location ranges were found to be significant to highly significant (i.e., p = .018 and p < .0005, respectively). The next most influential case types were CE4 case types followed by USO case types.

It is also worth mentioning that current distinctions between the above case types were made on the basis of where the case types were observed and the circumstances under which they were observed. The current names given to these types of experiences may, in fact, be a hindrance to an accurate understanding of their relationship, since all three case types share indistinguishable craft shapes. This was seen in Chapter 2 as craft shapes for all experiences were compared across three frequency categories: (1) Under 18 times; (2) (19-38) times; and (3) 39 times and two case types (i.e., a 2 x 3 matrix). The earlier significant results reported in Chapter 2 support the results reported now in Chapter 4.

Table 17: Cross Tab Results for the Relationship Between Case Types and Ranges of G. P. S. Range of Values Using Craft Shape Frequencies

Case Types	Ranges of G. P. S. Values				
	(2.0-4.0)	(4.1-6.1)	(6.2-8.2)	(8.3-15.0)	Total
CE4	96	14	8	16	134
UFO	151	42	8	11	212
USO	35	11	9	11	66
Total	282	67	25	38	412

Omnibus Chi Square Test

	Value	df	Significance
Pearson Chi-Square	24.016	6	.001

Effect Size (Moderate): Phi = .241; Cramer's V = .171; Significance = .001 (both measures).

Secondary Chi Square Testing

(1) CE4 (n=134) v UFO (n= 212) Case Types. UFOs may likely be involved in human abductions, since the frequencies of their craft sightings in the same locations as CE4 craft are highly significant across the four ranges of G.P. S. values presented below:

	(2.0-4.0)	(4.1-6.1)	(6.2-8.2)	(8.3-15.0)	Total
	96	14	8	16	134
	151	42	8	11	212

	Value	df	Significance
Pearson Chi-Square	10.102	3	.018

Effect size (Small): Phi = .171; Cramer's V = .171; Significance = .018 (both measures). The results favored UFO craft shape frequencies being mostly responsible for the significant results.

(2) CE4 (n = 134) v USO (n = 66) Case Types. The frequencies of sightings of USO craft shapes and reports of CE4 craft shapes *in the same G.P.S. location ranges* were not significant: χ^2 (3, N = 200) = 7.496, p = .058. Phi and Cramer's V (Effect sizes) were small (.194 for both) and both were insignificant (.058).

(3) UFO (n = 212) and USO (n = 66). G. P. S. location ranges and craft shape frequencies common to both UFO and USO cases were found to be highly significant: χ^2 (3, N = 278) = 19.137, p = .000 (p < .0005). The Effect size was medium (.262) for both Phi and Cramer's V and highly significant for both measures (p = .000; p < .0005).

Conclusions

1. Valuable insights were gained from determining relationships between the G.P.S. Range of Values for known abductions and the derived G.P.S. values of craft associated with UFO and USO events. These correlations were helpful in suggesting an explanation for why UFO and USO craft, while not specifically known to be involved in abductions, would have any correlation with craft shapes and locations already associated with abduction events. From Bivariate Correlations (Table 16) it was found that frequencies of *shapes from CE4, UFO, and USO events* produced very significant relationships.

2. It is *highly likely that UFOs are linked to abduction cases*, given their similarity in shapes and commonality in shape frequencies across location ranges with CE4 shape frequencies across the same four location ranges. Such craft shapes have been seen by observers (i.e., non-abductees) and described in accounts of abductees in the same location ranges.

3. The frequencies of sightings of USO craft shapes and their location data compared to reports of craft shape frequencies involved in abductions and their locations were not significantly related. Nevertheless, it is highly likely that both UFO and USO craft are also involved in human abductions. Craft shapes for these two types of phenomena are *indistinguishable from the shapes* abductees describe in CE4 cases. Moreover, to support this assertion, CE4, UFO, and USO craft shapes and their locations were examined in three analyses reported earlier in this research (i.e., see Chapter 2, pp. 67-68). The results indicated that the three sources of craft shapes produced significant relationships involving the three types of events. More data, however, from USO cases are needed to fully resolve these conflicting results.

4. The results pertaining to G. P. S. Location values of three groups of alien beings sorted over four G.P.S. location ranges cited in earlier findings show the dominance of the Humanoid alien species in abduction events compared

to other groupings of aliens (sometimes including Humanoids). That is, the results indicated that the G.P.S. Range of Values **(2.0-4.0)**, or (16-30)° N Latitude and (0)° W Longitude to (46-60)° N Latitude and (0)° W Longitude, was the range of locations where *Humanoids* (i.e., the grays) were observed by abductees working alone *65 times abducting humans*. Similarly, *Other Individual Aliens*, as a group, but no Humanoids, were observed conducting abductions *9 times*; and *Groups of 2 or more*, including Humanoids, were observed conducting abductions *35 times*. The Chi Square results were highly significant: χ^2 (3, N =168) = 147.286, p = .000 or p<.0005). All other G.P.S. Range of Values for the same groupings had Observed Ns smaller than their Expected Ns and did not influence the omnibus outcome in the first analysis. Thus, these findings establish a significant relationship between (1) different alien species observed during abductions (n =109/168 events) and (2) the G. P. S. locations where abductions occurred. The results all but eliminated chance as the best explanation for the results. Again, the results indicated that Humanoid beings (i.e., the grays), more so than other combinations of beings, were highly likely the abductors in various abduction events occurring at the (2.0-4.0) ranges of latitude and longitude described in this chapter.

5. The second highest frequency (n=30/168) for a G.P.S. Range of Values where alien beings most abducted humans involved the **(4.1-6.1)** range. In this scenario, the Humanoid species was observed abducting humans 22 times; Other Individual Aliens were observed 1 time; and Groups of 2 or more were observed 7 times working side by side Humanoids during abductions. The Chi Square result favoring the Humanoid alien group (only) versus the other two alien groupings in the **(4.1-6.1)** range of values was found to be: χ^2 (2, N = 59) = 10.263, p = .006. Results for the remaining two G. P. S. Ranges of Location Values and the three alien groupings was: χ^2 (1, N = 29) = 2.793, p = .095. Thus, alien beings across the G.P.S. Ranges [(6.2-8.2) and (8.3-15.0)] were equally likely to be involved in abducting humans in these locations. So far, much attention has been paid to analyzing data that produced highly likely findings involving (1) the identification of alien beings from experiences of abductees, (2) the frequency of G. P. S. locations of abductions correlated with

(3) the frequencies of shapes and transformed G. P. S. location data of UFO and USO sightings. In the next chapter, an analysis will be explored which addresses the frequency of states of mind of abductees (i.e., awake, sleep/dream sleep, and induced hypnosis) with respect to abduction events. The central issue explored was what experts had to say about abductees and, of course, what skeptics have often said. I believe the data and findings in Chapter 5 will set the record straight regarding the trustworthiness of the recollections of abductees.

An Analysis of States of Mind Among Abductees

There is not a single abduction case in my experience or that of other investigators that has turned out to have masked a history of sexual abuse or any other traumatic cause. According to Bryan, Mack stresses that "the reverse has frequently occurred—that an abduction has been revealed in cases investigated for sexual or other traumatic abuse." (Bryan, 1995, p.441)

It is worth reiterating that at the heart of the abduction phenomenon are issues relating to whether or not an abductee was awake, in some sort of dream/ sleep state...not fully awake..., or under hypnosis when discussing an abduction episode(s). Or, worse, can we trust their recollections? These areas of discussion engender questions that challenge whether abductions are real or if abductions are fabrications of the mind entirely or partly exacerbated by ideas implanted by zealous hypnotists. For example, in her book, How People Come to Believe They Were Kidnaped by Aliens, Susan Clancy offered a wide variety of arguments to address the causes of abductions, but other researchers strongly and convincingly argue that her work does not measure up to the task (Hopkins, 2005; Jacobs, 2006).

Are Abductees Reliable? The Experts Weigh In

David Jacobs (2006) provided several examples of why Susan Clancy's research on the abduction phenomenon lacked scientific rigor and he further clearly showed that her accounts of well-known sources of information exhibited factual flaws. For example, I currently own and years ago watched the *Bolero Shield* episode of the Outer Limits Series and agree with Jacobs's assessment of Clancy's lack of knowledge about the description of the alien being brought to earth on a laser beam shot into space by a scientist in his private laboratory. The being's description by Clancy was way off, as were her understandings of the abduction phenomenon overall. For example, compounding the Bolero Shield blunder are several examples of how Clancy failed to evaluate real phenomena by choosing to employ preconceived beliefs, instead of using readily available facts (i.e., she could have watched the episode as I and others did and would likely have remembered the being's correct description). Hopkins called her approach "faith-based" and not scientific (2005, p. 1 of 5). In the context of the scientific method, Susan Clancy is simply act two in a play where the opening act was played decades ago by Edward Condon who evaluated Project Blue Book's evidence with the same theory-dependent reasoning. That is, both persons apparently had their minds made up about the phenomenon they were investigating prior to scientifically evaluating readily available evidence (Hanshaw, 2004). Let's look now from another perspective. Findings from Chapter 3 of this research will be employed to address seven examples cited by Hopkins (2005, p. 4 of 4) as areas *not addressed or overlooked* by Clancy. The quotes from Hopkins follow his name and an item number, both in italics. Responses to the italicized information taken from an earlier chapter in High Probability (HP) may likely promote a clearer understanding of the abduction phenomenon without the veneer of theory-dependent thinking.

Hopkins, Item 1: She [hereafter understood to be Clancy; brackets my own] included no study of the patterns of well-known and clearly defined physical sequelae - scoop marks and straight - line cuts- that frequently appear on individuals after their abductions.

HP Response to Item 1: In Chapter 3, Consistency of the Narrative (Table 13, p. 73), findings from an analysis of the frequencies of specific *physical and psychological factors* described by abductees is presented. Clancy, like other skeptics, apparently chose to ignore these factors in favor of her own pre-conceived or ill-conceived ideas. Moreover, the perpetrators of the acts listed in Table 12 resulted from abductee identifications occurring in world-wide abduction events (i.e., *168 out of 216 cases*). These abductors were revealed in Table 2, p. 26, Chapter 1. The Chi Square results of the analysis of frequencies relating to the number of times various aliens were identified was found to be: χ^2 (5, N = 250) = 475.472, p = .000 (p < .0005). This means that the odds of these frequencies occurring by chance are less than 1 in 1,000. One could say, therefore, that it is highly likely that abductees saw who they described while aboard an alien craft. Additionally, the eight out of ten factors listed on p. (73) occurred at frequency levels that exceeded the frequency expected by chance: χ^2 (9, N = 1024) = 192.523, p = .000 (p < .0005). In other words, the chances of the seven physical and psychological factors occurring by chance was less than 1 in 1,000. Once again, such an outcome indicates that what abductees described as acts they endured at the hands of alien abductors was highly likely to be exactly what they stated. As for craft aboard which these events took place, see the reported statistics on p. 59. These findings indicate that the ships aboard which reported abduction events took place are highly likely to be what abductees described.

Hopkins, Item 2/HP Response: Item 2 (i.e., ground traces, altered soil, foliage, etc.) constituted variables that were not the focus of this research and were, therefore, not tracked or analyzed across cases involving close encounters of the fourth kind and their relationship to selected variables of UFO and USO sightings as identified in this research.

Hopkins, Item 3: She made no mention of the eye-witness testimony of neighbors observing a UFO over a house where an abduction is taking place; of witnesses who search in vain for an abducted child who is later found outside a fully locked house; of the incidents in which the police are summoned because of the temporary disappearance of a baby from his crib or a child from her bedroom, but who turn up, unobserved, an hour or so later; or hundreds of similar cases in which abductees are known to be inexplicably missing.

HP Response to Item 3: Presented throughout HP is the proposition that only one well-documented example is needed to contradict the proposed idea that aliens do not abduct humans or that no one other than an abductee has witnessed such an act. In fact, the Linda Cortile Napolitano case has elements of a craft being observed by *four* witnesses while an abduction took place. A similar event, though there are others, took place in Australia in "Tiaro, on (the) outskirts of central Queensland" (Chalker & Harrison, 2001, p. 1 of 4). The critical part of this incident is that the abduction of Amy Rylance was *witnessed* by a business partner, Petra Heller. Rylance, like Linda Cortile Napolitano, was seen by Heller being floated through the air by a square beam of light up to a waiting craft. Rylance, who was not conscious during her transport to the ship, was awakened while on board the ship where she found herself on a table with several beings surrounding her. Amy Rylance felt she had been on the ship for a while. She was returned by her captors, alone and in the dark, to a location some 790 kilometers from Tiaro. It is interesting that Amy Rylance explained this action by stating that the aliens told her the "lights were wrong" [at Tiaro] where they originally abducted her. Apparently, they noticed the presence of police lights at this location that were not there when they came the first time. After being returned at a different location, Mrs. Rylance made it to a service station where people there helped her to receive medical attention at a hospital and to contact her husband, Keith. Under penalty of prosecution for giving false statements, under Australian law, Amy Rylance signed a statement regarding the details of her abduction and return. Two cases, five witnesses. Obviously, every abductee is not crazy; but other evidence will be presented later in this section by experts to explain this assertion. The present information here, however, also will serve as a warning to critics that they should do their homework.

Hopkins, Item 4. She made no mention of the bizarre errors the UFO occupants often make, such as returning individuals from group abductions wearing someone else's clothes; replacing abductees in the wrong room or building after an abduction; or returning an individual to her bedroom in a locked and bolted house with her feet soiled and the back of her nightgown covered with damp leaves; or any of the scores of other such significant errors.

HP Response to Item 4: The Nobel Laureate, Kerry Mullis, says that prior to his abduction (see also, Chalker, 1999) in a wooded area near his cabin about 50 feet from an outside toilet, he saw a *talking raccoon*. The next thing he knew he was walking on a road back toward his cabin the next morning with no idea how he got there. This type of abduction illustrates both *screen memories* (i.e., the apparent ability of alien technology to project a life-like image of whatever they wish, instead of their own appearance) and *missing time*. Mullis could not account for the early morning walk back to his cabin on a road much farther from his original location that was near an outside toilet only 50 feet or so from his cabin. It was six hours after he originally left the cabin. Why was he not returned to the spot 50 feet from his cabin? How does a raccoon engage a human in conversation? Mullis was awake at the beginning of this ordeal and again at the end of this experience. One cannot avoid becoming wet outside walking around or standing in one place for six hours! Mullis indicated that his shoes/clothes were dry after six hours in the woods, even though dew was on everything else in those early morning hours returning to his cabin. Neither sleep paralysis nor implanted false memories are in play here (Hopkins, 2005), but missing time seems to be the phenomenon about which Clancy offers no counter argument; except to overlook six hours of critical missing time.

Hopkins, Item 5: She made no mention of the hundreds of cases in which two or more individuals are abducted at once, and whose traumatic memories match in every detail.

HP Response to Item 5: Both the Allagash abductions and the Barney and Betty Hill cases come to mind as examples of abductions involving multiple persons in a single event. Hypnosis was used by Benjamin Simon in the Hill case and by Raymond Fowler in the Allagash case. Both cases are part of the data base of this research. The Betty and Barney Hill experiences as well as the experiences of the five men in the Allagash collection of cases should not be thought of as cases about false memories just because the cases involved the use of hypnosis. Such reasoning by critics of abduction cases which involve hypnosis is flawed in that "the most appropriate conclusion that can be drawn from the available evidence is that hypnosis does not reliably produce more false memories than are produced in a variety of non-hypnotic situations in which misleading information is conveyed to participants (Lynn & Kirsch, 1996, p. 151). In relation to the reference to misleading information (i.e., critics often say that all hypnotists in abduction cases use leading questions to shape the abduction narrative), Lynn and Kirsch (1996) dispute the findings of Newman and Baumeister that the strength of the link between fantasy proneness and UFO abductions is not as solid as the afore-mentioned authors would suggest. Lynn and Kirsch point out that these authors "argue that fantasy proneness is correlated with hypnotizability and that people reporting abduction memories are highly hypnotizable; therefore, self-reported UFO abductees are relatively high in fantasy proneness" (1996, p. 153). Contrary to Neuman and Baumeister's claim, however, "the fact that a supposed memory was uncovered during hypnosis does not mean that the person who uncovered it is highly hypnotizable. The data on hypnosis and memory suggest that the production of new memories, accurate or not, is not much different with hypnosis than it is without hypnosis...Similarly, the relation between fantasy proneness and hypnotizability is quite small, with direct tests of the proposed links yielding negative or unimpressive results (e.g., Lynn & Rhue, 1988; Lynn, Rhue, Green, Mare, & Williams, 1992)" (1996, p. 153).

Hopkins, Item 6: She made no mention of a few accounts- such as the Travis Walton case or the Linda Cortile abduction - in which numerous witnesses see all or part of the abduction as it is being carried out.

HP Response to Item 6: In High Probability, both the Travis Walton and Linda Cortile (Napolitano) cases were evaluated to produce abduction scores for each of the 216 cases which, in turn, provided a reasonable idea of what close encounters of the fourth kind are like from the perspectives of abductees. In the Walton case, for example, Travis and five of his fellow loggers saw the space craft hovering above the ground just before the object shot a bluish-green beam that struck Walton knocking him several feet through the air before he hit the ground. No doubt there were witnesses who saw what Travis saw as he approached the craft before being struck. Frightened, all of his workmates left, but his foreman, Mike Rogers, soon returned to where Travis and the ship were located, but both were gone. Travis Walton, in a documentary presentation at the 2003 International Scientific & Metaphysical Symposium, (1) gave a detailed account of his time aboard an alien ship; (2) described the star-like background he saw as he pushed buttons and manipulated a lever on a chair he sat in; and (3) described other human-like beings (with non-human looking bright golden-hazel eyes) he saw and interacted with, as well as non-humans (greys) who examined him prior to releasing him ten miles from where he was abducted (Mill Creek Entertainment, 2011). Moreover, Walton also described other space craft he saw; some even larger than the disc shaped vehicle that appeared in the Apache-Sitgreaves National Forest in 1975. No debunkers have ever been successful at disproving any aspects of Travis Walton's story; not his phone call to his family 5 days after his abduction; not the lie-detector test he passed, or any explanation he gave regarding his abduction (Clark, 1998). Like Betty and Barney Hill before him, Travis Walton's abduction supports the contention that we are not alone. Linda Cortile Napolitano's case was made equally interesting by some of its key details. In this case, four witnesses saw this woman being floated through a *closed window* accompanied by four small grey beings and through the air in a bluish-white beam to a ship from her apartment building residence some 120 feet (12 stories) above ground. One of the four witnesses was former U. N. Secretary General Javier Perez de Cuellar; the other three were two of his bodyguards (on the ground by his stalled car) and a woman who was in her car on the Brooklyn Bridge at the time of the abduction. Linda sought and underwent hypnosis to then reveal what other witnesses saw and the details of her examination aboard ship and her later return to her bed where her sleeping husband lay just as he was prior to her being removed from her bed by the abducting beings (Hopkins, 1992). The Walton and Cortile (Napolitano) cases, among others, should put to rest the argument that there is no evidence of alien abductions of human beings.

Hopkins, Item 7. She made no effort to interview the friends and family members of the people in her sample, or, in fact, anyone who might have insight into their general trustworthiness and emotional soundness. Instead, Susan Clancy alone, because of her faith in the non-existence of UFO abductions, decided that all of her subjects' abduction accounts were false, and that all of their traumatic recollections were nothing more than false memories. She is, therefore, implying-indirectly but absolutely - that none of her subjects can tell the difference between dream and reality. To the public at large, this means, in effect, that an experimental psychologist with a Harvard degree believes everyone claiming UFO abduction experiences is suffering from a form of mental illness. For me, in the absence of any actual investigation of their accounts, such a radical, blanket condemnation by Susan Clancy of her innocent and naively trusting subjects is both ethically reprehensible and a disgrace to science.

HP Response to Item 7. In High Probability, an effort was made to build this research upon reliable information. It was not long before a search of the literature separated the signals of those willing to conduct an earnest search for truth through the use of the scientific method from the noise of those willing to use criticism, fear, and theory-dependent thinking to obfuscate the issues that surround a phenomenon such as close encounters of the fourth kind and associated events such as UFO and USO sightings. HP chose the science of statistics as a tool to analyze data in the manner explained in Chapter 1. Without much physical evidence, the evidence at hand had to be examined for patterns that could be represented or counted using event frequencies in order to present a picture of what people, honest everyday folks, were experiencing. Hence, to say that an event is occurring at a frequency greater than the frequency expected by chance is to state that *chance cannot explain* the outcome (s) being examined. The conclusion has to be, therefore, that some other explanation is more appropriate. In most of the results reported so far in this study, an overwhelming number of test results from the use of the Chi Square procedure indicates that the odds of events occurring due to chance are less than 1 in 1000 ($p < .0005$). If these were odds at a casino, most anyone wagering against the house would soon be rich beyond measure. This research, however, is about the lives of people that are being affected by a phenomenon that will not yield all of its secrets without an open-minded analysis of the patterns in data and the evidence made available to us, in part, *by those who have actually experienced* the phenomenon. It is highly likely that abductees are the physical and psychological sources of evidence in need of further study.

Conclusions

1. Many, if not most, ordinary persons who read about abduction events may wonder about the state of mind of persons who claim they have been abducted. Some of the reasons popular among persons who doubt the veracity of what abductees say include assumptions such as abductees must be crazy, or seeking publicity, or that they must have been dreaming or only partially awake or even suffering from some latent episode of sexual/physical abuse. Such explanations may bolster the belief system of some across the general public and, thereby, such reasons may help them grapple with the often-disturbing details of abductee experiences (Mack, 1994). Toward this end, three different states of mind (i.e., Awake, Dream/Sleep, and Hypnotized) were tracked across 216 abduction cases to gather frequencies in order to determine if these frequencies were greater than the frequency expected by chance. The omnibus test results were: χ^2 (2, N = 184) = 73.576, p = .000; p < .0005. The results were highly significant and greater than the frequency expected by chance. The results indicate that the odds that all three of these states of mind occurred due to chance was only one chance in 1,000. Thus, whatever abductees said they saw is highly likely to be what they described. A secondary test comparing the frequency of the Awake state (n = 95) to the frequency of the Hypnosis state (n= 82) revealed that the Awake state contributed most to the significant outcome, since the Hypnosis state's Observed frequency was smaller than its Expected frequency (n = 88.5). The same was true in the first analysis where the Dream/Sleep variable's Observed frequency (n =7) was less than its Expected frequency (n = 61.3). That is, χ^2 (1, N = 177) = .995, p = .328. These results mean that a person who is *awake* is no more (or less) likely to discern his or her abductor's identity than a person under hypnosis who may be asked by a hypnotist to identify or describe his or her abductor (s).

2. With respect to abduction cases judged to be due to false memories, such reasoning by critics of abduction cases which involve hypnosis is flawed in that "the most appropriate conclusion that can be drawn from the available evidence is that hypnosis does not reliably produce more false memories than

are produced in a variety of non-hypnotic situations in which misleading information is conveyed to participants (Lynn & Kirsch, 1996, p. 151).

3. With respect to the argument by critics that *self-reported abductees are high in fantasy proneness*, Lynn & Kirsch indicate that contrary to Neuman and Baumeister's claim, "the fact that a supposed memory was uncovered during hypnosis does not mean that the person who uncovered it is highly hypnotizable. The data on hypnosis and memory suggest that the production of new memories, accurate or not, is not much different with hypnosis than it is without hypnosis...Similarly, the relation between fantasy proneness and hypnotizability is quite small, with direct tests of the proposed links yielding negative or unimpressive results (e.g., Lynn & Rhue, 1988; Lynn, Rhue, Green, Mare, & Williams, 1992)" (1996, p. 153).

CHAPTER 6

# p < .0005 Means More Than You Think

In sum (and in practice), the researcher defines a level of risk that he or she is willing to take. If the results fall within the region that says, "This could not have occurred by chance alone — something else is going on," the researcher knows that the null hypothesis (which states an equality) is not the most attractive explanation for the observed outcomes. Instead, the research hypothesis (that there is an inequality or a difference) is the favored explanation. (Salkind, 2008, p. 157).

Periodically, I have reminded myself why this research has become more than just a strong desire to determine the next chi square value and its p-level of significance. Even though outcomes of events expressed in terms such as "not likely" or "highly likely" are important ways to support clarity in meaning for results, I was moved even more as I reflected on the fact that the statistical language used in this book is an attempt to peer through data sets in order to better understand reported abductions told in the accounts of ordinary human beings who say they have had a close encounter with beings who are not human. Hence, results such as p < .0005 (found often in this study) means that the probability that such events are due to chance is less than 1 in 1,000. This also means that some alternative explanation is more appropriate. Since abductions are supposedly nonexistent, not knowing how the results might turn out made me hold my breath. I asked myself questions such as (1) what is the comparison between the frequencies for six different alien species reportedly seen aboard space craft compared to the frequency expected by chance?; or (2) Do the frequencies for the number of men and women abducted across three age groups exceed the frequency expected by chance? After calculating Chi Square results from data related to these questions, I realized that it is highly likely that abductees have (1) seen the beings they described and (2) that, overall, the frequencies for men and women abducted world-wide across three age groups is greater than the frequency expected by chance. These realizations came about because in each case the significance of the calculated Chi Square values was the same: *p < .0005!* In other words, chance was not a likely explanation for the two questions investigated. Instead, the more likely explanations are (1) that the Humanoid species (grays) was most responsible for the omnibus outcome in question one and (2) the ≤ 30 age group was most responsible for the omnibus test results for question two. How is it possible to know this? It is possible because abductees (N = 216 cases) told what they observed and described their experiences to investigators. I then counted the number of times experiences of interest to me fell into certain categories. Also presented here in this chapter are the findings of experts in psychology that hypnosis is not as influential a factor in what abductees say or recall as critics contend and that some abductees were not asleep nor dreaming throughout their abductions. The abductees also had *one or more witnesses* who saw the abductions take place. The powerful combination of eyewitnesses to abductions, expert assessments about relationships between (1) *hypnosis and fantasy proneness;*(2) *hypnosis and false memories,* and (3) the *highly likely* statistical results (i.e., p < .0005) relating to how many times abductees saw six different alien species aboard a ship during their abductions, now must be considered as corroborating evidence that strengthens the relationship among these findings.

It is also clear that close encounters of the fourth kind are highly likely to be the experiences as described by abductees. With these findings in mind, several new questions in Chapter 6 were investigated regarding the frequencies for *Single or Multiple Abductions* and if they have a relationship to (1) Craft shapes described during CE4 events; (2) Geographical Regions; and (3) G. P. S. Ranges of Location Values. Questions were also examined about the relationship between all Case

Types in this study (i.e., CE4, UFO, and USO) and whether a relationship existed between the G.P.S. locations of the three case types sorted across four ranges of derived G.P.S. location values. In other words, are these case types occurring in the same locations world-wide? There were also secondary tests conducted to determine the most influential case type responsible for an omnibus result. Results for the various relationships examined were determined by using the Cross Tabs procedure for Chi Square (SPSS, 2011). These results are shown below in Tables 13 and 14. As you will see, the results are clearly in favor of events occurring at levels that eliminate chance as the best explanation for the various relationships examined. One could get the feeling that abductees and those who saw objects in the sky are trying to tell us different perspectives about the same phenomenon. If careful thought is given to what p < .0005 means, one might come away from the results in this chapter with more (truth) than one is willing to consider.

The results in Table 18, *Items 2 and 3*, indicate that there was not a significant relationship between the frequency of craft shapes reported by abductees during *single and multiple abductions* by G. P. S. Location ranges or geographical region. All results for craft shape frequencies x abduction type x G.P. S. location indicate that it is equally as likely that either craft shape type might take part in abducting individuals as any other craft shape. Abductions may occur in various locations and effect individuals no differently than people in groups. Craft shapes used are apparently due to the decisions made by the beings conducting the abductions and not if an abduction is single or multiple in nature.

In an examination of the relationship between case types (i.e., CE4, UFO, and USO) and G.P.S. location ranges, it was found that CE4, UFO, and USO cases share a highly significant relationship across the four G.P.S. ranges of location values. The omnibus result indicated that the (2.0-4.0) G.P.S. Range of values was the range of location values most influential in the overall result accounting for 282 of the 412 shapes reported across the three case types and four ranges analyzed: χ^2 (6, N = 412) = 24.016, p = .001. The Effect size of the relationship across location ranges was small (Cramer's V = .171) to moderate (Phi = .241) and highly significant for both measures (p = .001).

Also reported following Table 18 is an interesting comparison of *craft shape frequencies* across four *common ranges* of G. P. S. locations by *case type*. The interesting pattern shown in this analysis was developed by examining the rank of each Case Type determined by the frequency of craft shapes reported in each range of G. P. S. Values. By comparing the ranks of shape frequencies horizontally, the agreement between case types with respect to common craft shapes across G. P. S. ranges becomes evident. For example, the **(2.0-4.0)** range of location values shows the *highest frequency of craft shapes* across all case types from one location range to the next.

Table 18: Results for Relationships Between Single and Multiple Abductions and Three Selected Variables

(1) Craft Shapes

Abduction Type	(Disc; Domed Disc; Sat. Shape; Sphere)	(Cigar; Cylinder; Delta; Light)	(Other: Diamond; Egg; Oval; Rect. Football)	Total
Single	53	41	16	110
Multiple	60	11	11	82
Total	113	52	27	192

Chi Square Tests

	Value	df	Significance
Pearson Chi Square	14.901	2	.001

Effect Size (Moderate): Phi = .279; Cramer's V = .279; Significance = .001 (both measures).

(2) Geographical Regions

Abduction Type	North East	North West	South East	South West	Total
Single	18	74	13	9	114
Multiple	8	57	5	8	78
Total	26	131	18	17	192

Chi Square Tests

	Value	df	Significance
Pearson Chi Square	3.023	3	.338

Effect Size (Small): Phi = .125; Cramer's V = .125; Significance = .338 (both measures).

(3) G. P. S. Value Ranges

Abduction Type	(2.0-4.0)	(4.1-6.1)	(6.2-8.2)	(8.3-15.0)	Total
Single	71	21	4	18	114
Multiple	53	12	5	8	78
Total	124	33	9	26	192

Chi Square Tests

	Value	df	Significance
Pearson Chi Square	2.358	3	.502

Effect Size (Small): Phi = .111; Cramer's V = .111; Significance = .502 (both measures not significant).

Frequency of Craft Shapes Ranked Across Common
G. P. S. Ranges by Case Types

G.P.S. Range	CE4	UFO	USO	% Agreement
(2.0-4.0)	1st (96)	1st (151)	1st (35)	100
(4.1-6.1)	3rd (14)	2nd (42)	2nd (11)	66.66
(6.2-8.2)	4th (8)	4th (8)	4th (9)	100
(8.3-15.0)	2nd (16)	3rd (11)	2nd (11)	66.66
	N = 134	N = 212	N = 66	

This analysis shows that UFOs, CE4s, and USOs tend to appear in the same G. P.S. range of values: (2.0-4.0) which translates to (16-30)° N above the equator and (0)° from the PM to (46-60)° N above the equator and (0)°from the PM. The same consistency in craft shapes in a common range of G. P.S. values is also shown for locations heading from the North Pole toward the South pole as confirmed by the actual locations of cases in this study. The (6.2-8.2) range would include locations between the north pole and below the equator.

Conclusions

1. In a Cross Tabs analysis of *Craft Shapes by Single and Multiple Abductions*, the omnibus test results indicated that (Disc; Domed Disc; Saturn-shape; and Sphere), as a group, were most responsible for the Chi Square test outcome: χ^2 (2, N = 192) = 14.901, p = .001. The Effect size of this omnibus outcome was moderate (Phi/Cramer's V = .279) and significant (p = .001). Thus, Single abductions, more so than Multiple ones, involved the four shapes of craft listed above compared to other craft shapes (i.e., see Table 18).

2. In a Cross Tabs analysis of *Geographical Regions by Single and Multiple Abductions*, the omnibus test results indicated that on a regional basis, there was no significant relationship between Single versus Multiple Abductions occurring across the four geographical regions (i.e., NE, NW, SE, and SW): χ^2 (3, N =

192) = 3.023, p = .338. Earlier results, however, saw a significant relationship between regions based upon the total number of abductions occurring in all regions without distinguishing between Single versus Multiple types of abductions. These results support the conclusion that, overall, alone or with others, regardless of earth region, one is equally likely to get abducted.

3. *Single versus Multiple Abductions versus G. P. S. Ranges of Locations* were examined and there was no significant correlation in their frequencies across the four G. P. S. Location ranges examined: χ^2 (3, N = 192) = 2.358, p = .502. The Effect size was found to be small (Phi/Cramer's V = .111) and insignificant (p = .502). The conclusion here is that alone or with others, regardless of specific G.P.S. location, individuals are just as likely to be abducted as persons in groups. When type of abduction is ignored, as seen in earlier results, there is a significant relationship between abductions overall and the G.P.S. locations where they occur.

4. In an examination of the relationship between case types (i.e., CE4, UFO, and USO) and G.P.S. location ranges, it was found that CE4, UFO, and USO cases share a highly significant relationship across the four G.P.S. ranges of location values. The omnibus result indicated that the (2.0-4.0) G.P.S. Range of values was the range of location values most influential in the overall result accounting for 282 of the 412 shapes reported across the three case types and four ranges analyzed: χ^2 (6, N = 412) = 24.016, p = .001. The Effect size of the relationship across location ranges was small (Cramer's V = .171) to moderate (Phi = .241) and highly significant for both measures (p = .001).

5. The analysis of craft shapes (from primary and secondary testing) by G.P.S. location ranges across case types in this chapter revealed that UFO and USO sightings share a significant relationship with the known craft shapes and locations of CE4 events. It is, therefore, highly likely that UFO and USO craft shapes also may be involved in close encounters of the fourth kind, given the highly significant relationship and effect sizes they share with the craft shapes and locations of cases involving CE4 events (i.e., see Table 18, *Item 1* and the important findings regarding single v multiple abductions and craft shapes on p. 108).

CHAPTER 7

Other Sources of Evidence

Mc Donald had spoken before the AIAA sections at Vandenberg AFB in April and while there studied on-site the details of Case 35. As mentioned above, these radar-visual sightings occurred on the evening of October 6, 1967. An original, large object had been viewed for 45 minutes by a missile-range official at elevation 10-15° in the west-northwest. Unable to identify it, the official called another range official, who viewed it through binoculars. The large object was elliptical and the apparent size of "a large thumb tack." although it had red and green lights similar to aircraft, it was stationary for 45 minutes and was "fuzzy, like a spinning top." (James McDonald's Review of Case 35, Vandenberg AFB, October 6,1967; Druffel, 2003, p. 400).

Scattered throughout the literature are numerous examples of cases that destroy the argument that there is no physical evidence linking UFOs to intelligences that are not human. It is true that no alien bodies have ever been publicly displayed or examined to the satisfaction of parties on either side of the three-headed phenomena (UFOs, USOs or CE4s) investigated in this book. It is also true that there is no proof of human technology development capable of producing craft which have displayed flight characteristics that clearly indicate mastery of both gravity and inertial forces. Such an achievement would be needed to duplicate what has been observed and widely reported as recently as May 28, 2019 by two

U. S. navy pilots. More importantly, such a jump in technological development would also be needed to avoid killing human pilots. If it were true that humans have this level of technology, why are we still sending astronauts or cosmonauts into space using, by comparison, dangerously obsolete technology? An alternative conversation should focus on how to analyze all forms of data related not only to UFOs and USOs, but their apparent presence on this planet without an explanation of who is controlling what is being observed. Relative to this research, an answer is also needed to explain why such technology is involved in human abductions. A good start toward answers might well be to exhaustively study and analyze the accounts of people who have experienced such apparently non-human technology first-hand. The whole truth is, therefore, missing; at least, for now. The selected cases presented below and the types of evidence they involve do not agree with the convictions of those who say all UFOs and USOs are some type of natural phenomena (i.e., explainable by acts of man or nature or both). As far as evidence testable by science is concerned, all of us trained in the sciences know that the testing of evidence is constrained most by the evidence one is not willing to collect (Druffel (McDonald), 2003; Hill 1995). Like their CE4 companions, persons experiencing the behavior of UFOs or USOs have facts and evidence that strongly disagree with explanations that indicate that planets or mistaken identity or other explanations (anecdotal stories, crazy people, publicity seekers, etc.) are the best explanations for what people are experiencing world-wide. The selected cases from 1980 to 2000 are samples of why acts of nature or even supposed *secrets* of various governments) are not the best explanations for what happened in these cases. *Time will tell where reason won't.*

   1. **1980 Bentwaters-Rendlesham Forest Incident**. In the video, I Know What I Saw, directed and written by James Fox, (Gardner, Gardner, Frazer, and Craddock, 2009), actual notes, eyewitness testimony, and tape recordings made by U. S. Air Force personnel helped to reconstruct the now famous 1980 Bentwaters-Rendlesham Forest Incident. In his testimony at the National Press Club in Washington, DC on November 12, 2007, Sgt. James Penniston (Ret.) stated that during the night of December 26, 1980, over a three hour period, he and other U. S. Air Force personnel observed a craft maneuver about and then

land in the pine forests surrounding the base where he and he others worked. Of critical importance to this research is that Sgt. Penniston (1) got close enough to touch the craft; (2) made sketches of the craft and strange markings on the craft; (3) felt the electrostatic effects of being near the craft; and (4) he and others nearby watched the craft slowly ascend and then fly away, silently, in the blink of an eye (Gardner, et al, 2009; Kean, 2010). Radiological data indicated abnormally high background readings along with broken tree limbs in the area of the vertical path where the craft landed and later ascended back into the sky (see, for example, Col. James Halt's segment in the same video (Gardner, et al, 2009). Moreover, Sgt. Penniston later returned to the site and made cast impressions that, when connected by lines, formed a triangular shape just under "10 feet on each side" (Kean, 2010, p. 184). Sgt. Penniston also indicated that (1) "I put my hand on the craft, and it was warm to the touch. The surface was smooth, like glass, but it had the quality of metal, and I felt a constant low voltage running through my hand and moving into my forearm"; and (2) "I knew that this craft's technology was far, far, above what we could ever engineer" (Kean, 2010, pp. 180-181). While some have argued in favor of beacons (or planets in the night sky) as an explanation for what these airmen experienced, the details revealed by those present at the event provide solid reasons to doubt that the observed vehicles were man-made (i.e., two different vehicle types were seen during the two-night sighting). In other words, neither beacons nor planets break tree limbs, make soil indentations, emit beams of light at the feet of observers, or produce abnormally high radiological background readings (Gardner et al, 2009). The exercise of sound reasoning and attention to important details are still important components to any serious investigation of matters of this caliber. The variety of evidence in this case goes far beyond being viewed simply as *anecdotal evidence*, since (1) more than one person viewed this incident; (2) the evidence, including higher than normal radiological readings, plaster casts of soil indentations, broken tree limbs, and eye witness craft sketches support the overall narrative of the several witnesses to the event.

Also germane to UFO events is what persons educated in astronomy and engineering have to say about the level of technology that would have to be known to produce vehicles that perform like the ones seen by witnesses in the Rendlesham forest sighting and seen in other sightings as well. For example, Hill (1995, p. 311) wrote that "Educated people who accept the data of the UFO pattern at face value usually concede the probability that UFOs are produced by civilizations having at their disposal technologies far in advance of those available to man. The advanced technologies relate mainly to vehicle propulsive fields. Being knowledgeable of U. S. Government secrets on propulsion, I have known from the start that UFOs could not possibly be of earth-technology manufacture." Similarly, Alschuler, in a discussion about the performance of UFOs and their unlikely origins on this planet, says that "...in light of knowledge of inertia and acceleration, we can make several comments about possible alien technologies. First, if the reports of sudden changes of direction are taken at face value, then the UFOs, assuming they carry living passengers, must somehow have total control of inertia. Even if UFOs are only instrumented probes this would also likely have to be true, or they would have to be made of materials unknown to human science, because the inertial stresses they would undergo would destroy known materials and alloys" (2001, p. 42). Who or what is flying in our skies and performing maneuvers at speeds that would also kill human pilots?

2. **Japan Air Lines (JAL) Flight 1628.** This case involved a wealth of evidence consisting of eyewitness accounts, supporting radar evidence, and a meticulous analysis of the case by a federal agency head and an independent review by an optical physicist. At 5:10 p.m., local time, at 35,000 feet, Captain Kenju Terauchi, a pilot with 30 years of flight experience, along with his first officer and flight engineer, encountered a *mother ship* (and two smaller craft) over Alaskan skies (Clark, 1998). "His crew, Takanori Tamefuji and Yoshio Tsukuda, both saw it [the larger ship], too (Kean, 2010, p. 223). The three pilots described the UFOs in similar detail and performance characteristics. Captain Terauchi also drew an illustration of the larger craft and indicated that the two smaller craft approached his Boeing 747 cargo plane from between 100 and 500 feet, even though these craft were "about the same size as the body of a DC-8 jet (Maccabee, 1987)" Clark, 1998, p. 541). The larger craft and the smaller ones were tracked by an on-board radar, radar from the Regional Operational Control Center (ROCC), and Anchorage flight control, albeit at different times, but at least one other radar

system was tracking simultaneously with the on-board radar of JAL 1628 (Clark, 1998). Of key importance here are the following facts: (1) the "mother ship" and the other craft were seen by the crew and Captain Terauchi; (2) a sketch of the enormous craft was drawn by Terauchi who estimated its size to be that of "two aircraft carriers" (Clark, 1998, p.542); and (3) discrepancies about the event were clarified by an independent investigation by Maccabee (1987) who used a complete file of the incident compiled by the Federal Aviation Administration (FAA) to clear up confusion caused by skeptics who did not bother to wait for a complete report (Clark, 1998). John Callahan's National Press Club testimony comprised of visual, plane, and ground radar records supported what Captain Terauchi and his crew experienced (National Press Club, Washington, DC, November 12, 2007). Moreover, Clark indicated that Maccabee's analysis ruled out Jupiter (or Mars) because the crew indicted that the "arrays of lights rearranged themselves from one above the other to side by side, a reorientation that Jupiter and Mars would have found difficult to do" and because Jupiter was "at about the 10 o'clock position" [while] "the UFO "mother ship" was behind and to the left [of the 747] at the seven-to-eight o'clock position" (1998, p. 542).

3. **The Petit-Rechain Photograph.** During the period 1989-1991, a wave of sightings involving triangular craft occurred over Belgium in what has been called the Belgium wave (Kean, 1998). On an April night in 1990, the partner, of a lady walking her dog, used his camera to take the last of two color photographs remaining on the film. The one image that did materialize became the now famous Petit-Rechain photograph that a team of Belgian, American, and French experts eventually examined and found, among other findings, that the photograph "was not faked" due to many "unique characteristics of the lights" [on the craft] that would not be present "if the picture were a hoax" (Kean, 2010, p. 30). Additionally, one of the many experts, Professor Marc Acheroy, for example, "discovered that a triangular shape became visible when overexposing the slide" (Kean, 2010, p. 30). Another expert involved, Professor Andre Marion, "confirmed the previous findings" and, in his 2002 analysis using more sophisticated equipment, found that "the photograph revealed a halo"...and "...that within the halo, the light particles form a certain pattern around the craft like snowflakes in turbulence"

(Kean, 2010, p. 30). This snowflake pattern was hypothesized to be an indication of the craft's propulsion system; "magnetoplasmadynamic propulsion as suggested by studies done by Professor Auguste Meessen" (Kean, 2010, p. 30). The 1990 photo discussed here should not be confused with a faked photo taken on July 26, 2011 by a person known as Patrick M. who admitted as much to RTL, a Belgian TV channel (www.abovetopsecret.com/forum/thread805202/pg1).

4. **The Phoenix Lights**. In the video Special Presentation Secret Access: UFOs On the Record (Stern and Sudberg, 2011), Fife Symington, former Arizona Governor/military pilot, indicated that the craft he observed on March 13, 1997 was a massive triangular craft about 3,000-4,000 feet across that "made not a sound..." and "it had unbelievable aeronautical capabilities, something way beyond anything we have; that's for sure". This same craft was seen earlier by an estimated 10,000 people as it moved north to south across the state of Arizona beginning about 8:20 pm in Paulden, Arizona then south to Tucson, AZ and on to Phoenix, AZ (about 200 miles distance) until about 9:00 pm (Gardner et al, 2009). So, what was it that thousands of people saw over parts of Phoenix, AZ for over 106 minutes? The Phoenix Lights case demonstrates that eyewitness accounts are valuable and reliable sources of information given the consistency of what was reported and described in detail by such a diverse group of eyewitnesses (Kean, 2011). A sighting by 10,000 people does not fall under the definition of anecdotal evidence. Look it up. This case appears, instead, to fall under the heading of finding out what one wishes to find out. An object that large flying that low over a major population center and within its airport space is a violation of federal safety regulations; so, my reaction here is hardly overstated. One should fact check these circumstances as well.

5. **The Illinois UFO, January 5, 2000.** In the video The Edge of Reality: Illinois UFO, January 5, 2000 (Barker, 2000), a truck driver, Melvern Noll, reported a huge object "as big as a two story house" moving over Highland, IL. He indicated that "if you would not have been looking up, you would not [have] known it was going over you." One of the three police officers, Officer Barton, who saw the object during the time it moved over his location, in Lebanon, IL, said a large triangular vehicle came toward him at an altitude of about 1000 ft,

stopped, and made a flat turn (rotated in one spot) before it moved away in "2-3 seconds, tops" to Shiloh, IL "about six miles away". Less than a minute later, Officer Martin, In Shiloh, IL, had the object in sight and said it was a triangular or wedged-shaped object with a base about as long as a football field. The underside of the object resembled something "put together in pieces"... "not smooth, but like plumbing" on the underside. Later, another policeman, Officer Stevens, reported to the dispatcher that he also had the "arrow-headed shaped" object in sight over Millstadt, IL. He said "its huge" and that the object had three lights; one red light in the center and one white light at each corner of a triangular shape that also had a concave, white strobing bank of lights across the rear (Barker, 2002, p. 11 of 38). By compiling all eyewitness descriptions, police transmission tapes, and computer reconstruction technology, the object was determined to be longer and larger than a 747 jet liner or a B-2 bomber, and it moved at an estimated speed of 3600 miles per hour (Barker, 2000). The silent movement, and at times hovering ability, of such a large craft and the effects of inertial forces estimated in the tens of hundreds, rules out earth-based vehicles with human pilots and man-made materials due to the forces on man-made materials and pilots for reasons cited earlier in this paper. Moreover, Putoff (1998), commenting on the state of advanced propulsion technology (i.e., technology based upon the polarizable vacuum theory), stated that "unfortunately, making such requirements [engineering requirements] is beyond technological reach without some unforeseen breakthrough (Pfennimg & Ford, 1997)" (p.6).

Hence, the presence of bright light sources without the normally expected presence of flames from discernable engines; the "blink-of-an-eye" aeronautical capability; the absence of jet engine noise; the apparent material science and propulsion advancements evident in the ability of such vehicles to accelerate to unknown speeds, and the apparent control of inertial forces makes the Rendelsham Forest, JAL Flight 1628, and the Petit-Rechain cases *unlikely* examples that are representative of vehicles made by man. Additional cases in this chapter will be additional unlikely examples. Moreover, to attribute such advancements to super-secretive government-financed programs (or to private/public ones, for that matter), does not pass the *known science* or *revealed principles of science* smell test.

For example, who got the Nobel Prize for figuring out how to control inertia and gravity? That's what I thought. From findings in this study, I offer the following: If our government (or some other entity) is responsible for the funding and/or development of such vehicles discussed in this research, then I would require that *physical evidence* be provided in order to give credit to human science and engineering for such an outstanding technological and material science achievement. "From the beginning, the biggest obstacle to scientific research into the UFO problem has been the absence–or at least the paucity–of physical evidence" (Sturrock, 1999, p. 162). But let me be clear regarding the requirement of physical evidence. Ownership of the kind of technology mentioned in the above examples and others to follow requires that such an agency or government be able to produce the physical evidence on-demand (i.e., a pre-planned landing at O'Hare International will do). In this way, producing physical evidence to legitimize an "extraordinary claim" will then, logically, apply to any government and/or private or public agency just as it does to ufologists or to others who may believe the technology seen flying in our skies and abducting humans is not attributable to any efforts by man alone.

Conclusions

1. Each of the five historical cases presented above give clear examples of incidents that offer the following types of physical evidence that raises serious doubt that man is alone in the universe: (1) radar-visual; (2) photograph of a large triangular craft determined by experts not to be a fake (3) plaster casts of landing gear ground impressions that also imply the weight of the craft that made the impressions; (4) a sketch of a craft touched by an airman investigating an incident; (5) thousands of eyewitnesses observe a large triangular craft as it traveled from one end of a state to the other; and (6) in a separate account years apart from the other triangular craft sighting, multiple policemen observe a large triangular craft and their synchronized comments and radio logs tell a credible and detailed story.

2. The five cases cited in this chapter leave little doubt that physical evidence is in-hand and has been available to researchers and the scientific community

over recent history (1980s to 2000). Earlier than this time frame, however, are even earlier accounts (Hanshaw, 2004) that include the following types of evidence: *radar-visual* (Washington Merry-Go-Round); findings from a *word analysis* of the Ramey Memo (Rudiak Analysis); *chemical trace evidence* (Ted Phillips Soil Research); and an astronomical *star map* vetted years later by an astronomer, Marjorie Fitch (Betty Hill Star Map; the *stars* in the map *were not visible from earth* in 1961).

3. We do not know how or why the craft seen in our skies behave as they do or how they are able to come out of or fly into our waterways as easily as they do. But they do. If CE4, UFO, and USO cases are considered based upon the evidence at hand, the probability of arriving at a reasonable answer becomes convincingly clear; even if the *box of answers* come wrapped in a box that says: the answers within have a very, very, *small chance* of not being correct.

CHAPTER 8

The Research Conundrum

The subsections below are issues for both thoughtful reflection and future research. The path of choice is challenging, given the stakes involved. The path I hope others will take may not go as far as writing a book, but certainly I hope the nature of the phenomena discussed here will motivate others to read more and make informed decisions about the nature of CE4, UFO, and USO events. Regardless of what is decided, the desire to know the truth about the unknown is a faithful lover. Pursue the truth and the rewards will justify themselves.

Framing Findings about the CE4 Phenomenon and Related Phenomena

The largest portion of my task in this book was to frame the results of statistical testing of hypotheses about CE4 experiences in such a way that readers might finish a sentence or paragraph containing somewhat unfamiliar symbols and expressions, but still be lead to appreciate a particular finding because the results were restated as either *unlikely (p > .05)* or *highly likely (p < .0005)* outcomes (i.e., recall that the symbol (<) means *less than* and its companion (>) means *greater than*). Readers will find test results stated this way through out High Probability. The results were predominantly more near the *highly likely* end of the spectrum than the unlikely end. This basically means that for most tests conducted, *chance was not the best explanation* for a given hypothesis tested. Some other explanation was better (i.e., results such as p = .008 indicate that it is very likely that alien beings preferred to communicate telepathically as opposed to other reported forms of communication, for example). In addition to statistical results, there were insights from experts in appropriate fields whose findings dealt with issues critically important to the abduction phenomenon. For example, it was made clear that hypnosis is not as unreliable as critics have stated. Moreover, as pointed out in Chapter 5, for example, theory-dependent reasoning is not the mind set of choice in any attempt to solve a problem in a scientific investigation. Such reasoning distorts the facts and obfuscates understandings across interdisciplinary aspects of an already complex phenomenon. In this research, a series of questions were explored, and answers were presented from an analysis of data collected as close as possible to the sources of the experiences under examination. Hence, it was paramount to determine from the research of experts in appropriate fields if the experiences of abductees should be relied upon as viable sources of trustworthy information. Again, Chapter 5 and additional references in this book accomplished that task. With reliable information in hand, the remaining task was to research multiple accounts of the same abduction event(s) prior to inclusion into the data set and rely upon the expertise of law enforcement agencies, investigative reporters, and people and organizations who have a desire to determine the truth of what happened; even if doing so stresses their personal beliefs. The results of such investigations and determinations and the application of acceptable procedures for analyzing frequency-based data enabled this research.

The internet and a couple of thousand dollars in books by authors whose works are standard bearers, with respect to CE4, UFO, and USO events, heavily impacted the choice of works relied upon to clarify the complexities of cases used in this research. A brief tour of the reference section of this book will make it clear who some of those experts are. Many other works helped to shape my understanding of relationships between CE4, UFO and USO experiences. Variables such as dates, ages of CE4 witnesses, and G.P.S. related data, etc. missing from some reports eliminated what was otherwise very informative information in the books of some authors. The rest of the work was left to me. Reports were read and methods developed in this research which facilitated the development of case information (i.e., the adaptation of Bullard's Eight Episodes in Chapter 1, for example). The procedure which transformed regular G. P. S. information into a form which made it easier to correlate the frequency of locations with the frequency of other variables was an insight I found to be rewarding, since it produced results that agreed with the findings of others which were arrived at by other means. This approach also helped to produce *highly likely* findings not previously known; at least not at the level of significance presented in this study.

Research of this nature still lacks large volumes of physical evidence; some is available, however, as presented in Table 12 (Items 1 and 4), Table 13, Table 14 and the variety of physical evidence mentioned in the selected cases in Chapter 7. This is not enough physical evidence for a slam dunk case, by any metric. Nevertheless, the cases analyzed in HP show that it is not zero physical evidence for the existence of alien beings and their craft either. What is obvious is that more investigations and evidence are needed. More evidence from fields such as psychology and psychiatry could add clarity to the discussion of abduction experiences in ways that better inform researchers interested in solving problems germane to the abduction phenomenon.

Nevertheless, the conundrum of highest complexity is the ambivalent treatment of the proverbial eight hundred pound gorilla in the room: the *physical and psychological evidence* available in the bodies and minds of *abductees*. I suppose James McDonald had a point when he indicated that if a system of finding information is set up in such a way as to overlook that which one claims to be looking for, the results will confirm what supposedly was being sought in the first place (Druffel, 2003). The CE4, UFO, and USO experiences of ordinary people appear to be still suffering from this way of thinking and acting.

Consider, for example, the following sub-sections that beg for answers from modern science and other disciplines. These sections point out how much is yet not clear not only about CE4, UFO, and USO experiences, but the kinds of mysteries that seem to strongly imply that perhaps the history of mankind is not solely a history shaped by actions of humans alone. In other words, is it possible that early man could have been visited by a more advanced, space-faring civilization that shared some of its knowledge and then moved on? Read the accounts below, first. Then, after each section below, ask yourself is it possible? Perhaps thinking about questions in this way might reveal how much more might be gained if some of the older mysteries important to science, history, and evolution got the attention they deserve. Readers may agree that older explanations are proving themselves to be inadequate as inferred by questions raised in the subsections below. Without a complete resolution of valid questions about mankind's history and achievements, the whole truth will remain unknown.

Astronauts and Abductees

Is there a relationship between the experiences of astronauts during space travel and what abductees say they experience? The human body is affected by extended periods of time spent in space. If the medical community could be more engaged in CE4 research, perhaps data bases world-wide could be developed to measure how similar or different levels of acetone might vary in a comparison of time spent in space by astronauts compared to time spent in space by those who claim to be abductees. Perhaps space faring countries could share data about their astronauts that could, in turn, shed light on how long abductees may have been in space, given the human body's reaction to prolonged weightlessness.

Are Abductions a Public Health Issue?

When humans get abducted by other humans, such acts get condemned in most civilized societies and, once tried and convicted, perpetrators of such acts get imprisoned. I see no reason to give beings who are not human a pass simply because they are not human or because they may not be familiar with our laws and customs. What do the abductors want from humans anyway, since they are thought to be so much more advanced? Why can't alien beings simply ask the people of earth for what they seem so willing to take without permission? Abductees tell us in overwhelming numbers that these beings have an agenda which does not include a sensitivity to questions to which humans might desire answers. Why do we castigate abductees for suffering a fate most have revealed they never wanted but have been forced to experience; sometimes more than once, or worse, from childhood to adulthood. Why the double standard? Will abductions lessen if more investigations are vigorously conducted? Alien abductions are estimated to occur many thousands of times more than those already known (Hopkins, Jacobs, and Westrum, 1992). Even if such estimates turn out to be flawed as some have indicated (Blackmore, 1998), half of the estimated 4 million people abducted would still be a staggering figure to contemplate [i.e., (4 million ÷ 2) = 2 million; and 2 million abductions ÷ 50 states = 40,000 people per state!!). From another perspective, is it possible that currently unknown abductees might be encouraged to step forward and tell their stories, if the abduction phenomenon were treated as a public health issue? Readers should recall that the probabilities shown in this research are heavily in favor of abductions being literal events and not events due to chance. The next big challenge in abduction research is to improve techniques of surveying to eliminate as much as possible flaws in identifying abductees while improving incentives that might ethically encourage their participation in abduction research.

Ancient Structures and Ancient Objects that Defy Explanations for How They Were Built

1. *Puma Punku,* near Tiahuanaco [Bolivia]. Who built this site and how could massive stones have been cut and moved into place by whatever means that existed at that time? Quarries nearest to this site are over 20 miles away (Childress, 2000). Whatever was done then still is difficult to accomplish using even today's heavy equipment cranes and laser cutting equipment. How were megalithic stone blocks that weighed an estimated 150-200 tons or more cut, moved, and lifted by members of a culture preceding the invention of the wheel? How did these ancient people accomplish this? The implications are clear, if not done by human thought or technology, then to whom does such credit belong? And what is the evidence for that?

2. *The Saccara Bird, circa 3rd Century, B. C., Saccara, Egypt.* Found in a cave, this wooden image of what looks like a bird was later examined by two aerospace engineers, Eenboom and Apel (History Channel, 2010). They determined the bird-like figure to be a glider. Aviation and dynamics expert Simon Sanderson

later built a replica model of the glider, added a missing piece (an elevator that was broken off which birds do not have). He also wind tunnel-tested the model. Overall, they proved the glider could actually fly (A & E Television Networks, LLC, Ancient Alien Series, History Channel, History Channel, 2010). Was this a stroke of luck that someone in Egypt in the 3rd century B. C. studied the flight of birds and then built a wooden model that differs from birds and yet flies? How did they know a rudder of a specific shape would be needed to make the model fly? Was this a lucky guess? What are the odds of that? I wonder why persons of that time did not make drawings of this engineering achievement on walls inside of pyramids or etched such shapes in obelisks in other locations of Egypt like they did with other important things in their lives?

3. *The Gold Flyer: South America, circa 300-1600 A.D.* The *Gold Flyer* was found in a grave site of the Tolima people near present day Colombia. The swept-wing, gold figurines were examined and found to be more like modern jet fighters than any insect or birds ever known. Like the Saccara bird, aerospace engineers (i.e., Eegenboom and Belting, History Channel, 2010) built a larger scale model with an engine in 1997. The object was remote controlled and flew perfectly. Where would people of that era get such an aerodynamic design idea? Did they see something in the sky and make a model of it using their jewel-making skills? How close was the object and for how long was it present to get the level of detail so precise in the model found? Why did the level of detail include a fuselage, cockpit, delta swept wings, and stabilizers (i.e., birds and insects don't have stabilizers). A pattern is clear with respect to man's curiosity about ancient technology. What is suggested here is that, eventually, man will do to the detailed drawings of Vimanas mentioned in accounts of ancient texts in India what has already been done to the Saccara Bird and the Gold Flyer. Will mankind be curious enough about ancient technology to revisit descriptions of what supposedly existed thousands of years ago? Who will be the next people on earth to elevate humankind's understanding of aerospace engineering and propulsion principles to levels not yet achieved?

Are Alien Beings Pursuing a Peaceful Agenda?

My reasoning says that alleged alien beings have had ample opportunity to communicate with humans prior to (or after) landing on this planet. The reasonable grounds for this view is the behavior of these beings during abductions when they communicate telepathically with humans and tell them not to be afraid. If these beings can say what their intentions are in plain English during an abduction, why can't they just fly around the planet and broad cast their message electronically without landing? Their technology meshes with ours well enough to disable nuclear missiles in silos that are miles apart (i.e., the Minot Air Force Base Minuteman Missile Incident in 1967). This would be a far less traumatic way to get to know humanity. I am confident alien beings could, if desired, communicate their intentions electronically in ways that humans could understand, if they wished to communicate their real intentions. For the sake of argument, let's say that aliens have trouble communicating in our language by radio waves. Why not give their message (telepathically) of *no harm intended* to any abductee on any continent and let them deliver the message *along with an artifact* from their ship or planet as a sign of good faith? Appropriate authorities and the media might likely see such an act as a show of good will. Remember, as well, what happened in the Betty and Barney Hill case as Betty was leaving with a book one alien gave her and then this act was reversed by other beings on board the ship who objected. I wonder why such viable alternatives have not been tried more often. Instead, these beings covertly move about the planet, mostly at night, and abduct humans; in most cases, against their will. Are such actions those of a peaceful agenda?

Where to go From Here

It is my hope that the impact of this book will be to produce clearer perspectives about the abduction phenomenon and its relationship to UFOs and USOs. The conclusions presented here were based upon statistical test results that often stated that chance was not the best explanation for every hypothesis examined. For example, one such finding was that it is highly likely that a number of alien beings identified by abductees tended to work harder by themselves than when working with other aliens. Disc shaped alien craft dominated the shapes that are highly likely to be involved in an abduction event. Sightings of UFOs and USOs in locations involving CE4 encounters may offer an opportunity to evaluate the predictive efficiency of the prediction equation developed in this research (i.e., see p. 45). Perhaps a quality monitoring system in selected regions can document sightings in areas predicted by the equation. Moreover, the study of close encounters of the fourth kind is as valuable, if not more so, than the study of UFOs and/or USOs. The reason is simple. What is more valuable than a trustworthy eyewitness or a group of such witnesses? Abductees are the only eyewitnesses to CE4 events (in most cases) and evidence has been presented that they are no more likely to be fantasy prone, for example, than non-abductees. This book analyzes, first and foremost, the experiences of abductees and emphasizes, as well, a silent but important third party: the alien beings discussed here in, as well. This aspect of CE4 experiences cannot be overlooked or dismissed. The numbers say the same. Relative to these interrelationships, consider the viewpoint of a person who is both an abductee and a researcher:

In spite of what some prominent abduction theorists tell us about avoiding thinking in terms of "good and evil" or "positive and negative" when it comes to the aliens, this simply cannot be done, nor should it be. For these women, for my husband and myself, for all abductees, knowing that we have been made a part of this agenda and that we have been implanted, trained, and programmed to participate in some future scenarios, how can we not ask to what purpose our minds, bodies, and souls will be used? How can we put aside our rationality, our learned wisdom, and our ethics to trust the words and actions of beings whose nature is kept hidden from us and whose agenda involves the entire world? (Karla Turner, 2013, p. 243)

I hope readers will gain many informed perspectives from reading the research presented in **High Probability**. Let me know what you think: lhanshaw@ bellsouth.net. Thanks, in advance, and for buying this book.

References

References marked with an asterisk indicate cases used in the scoring process that met selection criteria.

*1987: Abduction of Jason Andrews. Retrieved from http://www.ufocasebook.com/jasonandrews.html

A & E Television Networks, LLC (Producer). 2010. Ancient aliens: Season one [The Evidence: Saccara bird and Gold flyer segments]. History Channel. Available from History.com.

A & E Television Networks, LLC (Producer). 2010. Ancient aliens: Season four [The Mystery of Puma Punku segment]. History Channel. Available from History. com.

*Allan, B. (2000). The A70 abduction case [Gary Woods case]. Retrieved from http://www.ufocasebook.com/a70abduction.html

*Allan, B. (2000). The A70 abduction case [Colin Wright case. Retrieved from http://www.ufocasebook.com/a70abduction.html

Alschuler, W. R. (2001). The science of UFOs: An astronomer examines the technology of alien spacecraft, how they travel, and the aliens who pilot them. H. Zimmerman (Ed.) New York, NY: St. Martin's Press.

Appelle, S. (1995/96). The abduction experience: A critical evaluation of theory and evidence. Journal of UFO Studies, n. s. 6, 29-78.

*Aston, W. (1997). Udo Wartena contact case in 1940. Retrieved from http://www.ufoevidence.org./cases/case1032.html

Barker, D. (Producer and Director) (William Matthews and Darryl Barker Photographers) (2000). The edge of reality: Illinois UFO, January 5, 2000. [DVD]. Available from Darryl Barker Productions at http://dbarkertv.com

Barker, D. (January 18, 2012). January 5, 2000, St. Clair CO., Illinois flying triangle case analysis. Retrieved from http://dbarkertv.com/UPDATE.html

*Basterfield, K. (2005). [1930's female case]. Retrieved from http://www.project 1947. com/kbabduc0505.html

*Basterfield, K. (2005). 1956 Miss L. case. Retrieved from 1947.com/kbcat/ kbabduc0505.html

*Basterfield, K. (2005). 1957/58 Sandy's case. Retrieved from 1947.com/kbcat/ kbabduc0505.html

*Basterfield, K. (2005). 1969 Susan's case. Retrieved from 1947.com/kbcat/ kbabduc0505.html

*Basterfield, K. (2005). 1961 Tasmania [Wayne's case]. Retrieved from http://www. casebook.com

*Basterfield, K. (2005). 1962 New South Wales [3 adults case]. Retrieved from http:// www.casebook.com

*Basterfield, K. (2005). 1967 New South Wales female case. Retrieved from 1947. com/kbcat/kbabduc0505.html

*Basterfield, K. (2005). 1966-67 New South Wales male case. Retrieved from http:// www.project 1947.com/kbcat/kbabduc0505.html

*Basterfield, K., Elsbeth, J., & Jones, P. (2005). 1970 Sydney, New South Wales male case. Retrieved from http://www.project 1947.com/kbcat/kbabduc0505.html

*Basterfield, K. (2005). 1972 Largs Bay, SA [Carol's case]. Retrieved from http:// www.casebook.com

*Basterfield, K. (2005). 1973 Winifred C. case. Retrieved from http://www.project 1947.com/kbcat/kbabduc0505.html

*Basterfield, K. (2005). Hay Plains, New South Wales case. Retrieved from http:// www.casebook.com

*Basterfield, K. (2005). 1979 Jindabyne, New South Wales case. Retrieved from http://www.casebook.com

*Basterfield, K. (2005). 1982 Queensland female case. Retrieved from http://www. project 1947.com/kbcat/kbabduc0505.html

*Basterfield, K. (2005). 1987 Brisbane female case. Retrieved from http://www.project 1947.com/kbcat/kbabduc0505.html

*Basterfield, K. 92005). 1991 New South Wales female case. Retrieved from http:// www.project 1947.com/kbcat/kbabduc0505.html

*Basterfield, K. (2005). 1993 Geelong, Victoria female case. Retrieved from http:// www.project 1947.com/kbcat/kbabduc0505.html

*Basterfield, K. (2005). 1996 Scott Longley case. Retrieved from http://www.project 1947.com/kbcat/kbabduc0505.html

*Basterfield, K. (2005). Susan case. Retrieved from http://www.project 1947.com/ kbcat/kbabduc0505.html

*Basterfield, K. (2005). 1996 Wendy Longley case. Retrieved from http://www.project 1947.com/kbcat/kbabduc0505.html

*Basterfield, K. (2005). 1996 Western Australia [Irene Sander case]. Retrieved from http://www.casebook.com

*Basterfield, K. (2005). 1996 Carol Nagel case. Retrieved from http://www.project 1947.com/kbcat/kbabduc0505.html

*Basterfield, K. (2005). 2001Sydney, New South Wales female case. Retrieved from http://www.project 1947.com/kbcat/kbabduc0505.html

*Before It's News (2014). One of the strangest alien abduction cases in China ever told [Meng Zhaoguo case]. Retrieved from http://beforeitsnews.com/ paranormal/2014/07/one-of-the-strangest-alien-abduction-cases-in-...

*Birdsall, G., Callaghan, R., Leir, R., Streiber, A., & Streiber, W. (1999). Alien abduction: A never-heard-before-encounter from Central America [Salma & family case]. Retrieved from http://www.mercuryrapids.co.uk/articles/ ALIENABDUCTIONAneverheardbeforeencounter...

*Blomqvist, H. (January-December,1986). An abduction in Sweden? Archives for UFO Research Newsletter, Issue (29),1.

*Booth, B. J. (1994). Infant disappears in Killeen, Texas. Retrieved from http://www. ufocasebook.com

*Booth, B. J. (2001). The salter encounter. Retrieved from http://www.ufocasebook. com

*Booth, B. J. (2007). The Ilkley Moor alien. Retrieved from http://www.ufocasebook. com/ilkleymoor.html

*Booth, B. J. (2008). The Fayetteville, NC encounter, January 8, 2007. Retrieved from http:www.ufocasebook.com/2008c/fayettevillenc.html

*Booth, B. J. (2011). Buff Ledge abduction [Michael Lapp case]. Retrieved from http://www.ufocasebook.com/Buffledge.html

*Booth, B. J. (2014). Woman suffers two alien abductions. Retrieved from http:// www.ufos.about.com/od/aliensalienabduction/fl/Woman-Suffers-Two-Alien- Abductions.html

*Booth, B. J. (2015). 1966-Encounter at Gulfport, Mississippi Eve's case]. Retrieved from http://ufos.about.com/od/bestufocasefiles/p/eve.html

*Booth, B. J. (2015). 1966 Hunters alien abduction in North Bay, Ontario Forest. Retrieved from http://ufocasebook.com

*Booth, B. J. (2016). 1968 Buff Ledge abduction. Retrieved from http://www. ufocasebook.com/Buffledge.html

*Booth, B. J. (2015). 1975 Abduction of Charles L. Moody. Retrieved from http:// www.ufocasebook.com/moody.html

*Booth, B. J. (2015). Danger down under: The Christa Tilton story. Retrieved from http://www.ufocasebook.com/christatilton.html

*Booth, B. J. (2015). 1976 Stanford, Kentucky abductions. Retrieved from http://www.ufocasebook.com./Stanford.html

*Booth, B. J. (2015). 1979 Dechmont forest abduction Robert Taylor case]. Retrieved from http://www.ufocasebook.com/taylor1979.html

*Booth, B. J. (2015). 1983 Alien abduction in Mobile, Alabama [Pat Norris case]. http://ufos.about.com/od/aliensalienabduction/flAlien-Abduction-near Mobile-Alabama...

*Booth, B. J. (2015). Mid-1980's: Unexplained events of abduction and lost time-Long Island, New York. Retrieved from http://www.ufocasebook.com/2008c/losttimenewyork.html

*Booth, B. J. (2015). 1988 Abduction of Bonnie Jean Hamilton. Retrieved from http://www.ufocasebook.com/moody.html

*Booth, B. J. (2015). 1993 Kelly Cahill encounter. Retrieved from http://www.ufocasebook.com/Cahill.html

*Brookesmith, P. (1998). Alien abductions [Myrna Hansen case] (p.32). New York, NY: Barnes & Noble.

*Brookesmith, P. (1998). Alien abductions [Clark Hathaway case] (pp. 134-135. New York, NY: Barnes & Noble.

*Brookesmith, P. (1998). Alien abductions [Kathy's case] (pp. 32-40). New York, NY: Barnes & Noble.

*Brookesmith, P. (1998). Alien abductions [John Velez case] (pp. 112-113). New York, NY: Barnes & Noble.

*Brookesmith, P. (1998). Alien abductions [Jack & Peter Wilson case] (pp. 86-91). New York, NY: Barnes & Noble.

*Bruno, E. (2012). Brazil: the Barroso case (1976). Retrieved from http://www.ufodigest.com/eliasBruno

Bryan, C. D. B. (1995). Close encounters of the fourth kind: Alien abduction, UFOs, and the conference at M. I. T. New York, NY: Knopt.

Bullard, T. E. (1998). The abduction phenomenon. In Jerome Clark (Ed.) the UFO encyclopedia: the phenomenon from the beginning (Vol 1: A-K, pp. 1-26). Detroit, MI: Omnigraphics.

Bullard, T. E. (2010). The myth and mystery of UFOs. Lawrence, Kansas: University Press of Kansas.

*Campbell, C. (2009). Utahns abducted by aliens. Retrieved from http://www.cityweekly.net/utah/utahns-abducted-by-aliens/Content?Mode=print&oid=2140...

*Chambers, H. V. (1967). UFOs for the millions [Daniel Fry case] (pp. 124-128). Los Angeles, CA: Sherbourne Press, Inc.

Cherniack, D. (Producer). 2010. UFOs: the secret history [DVD]. (Bernard Haisch, babes in a cradle segment). UFOTV, The Spirit Culture Preservation Project, and All in One Films (Distributors). (Available from UFOTV, 2321 Abbot Kinney Blvd, Venice, CA 90291).

*Chalker, B. (1976). Eliot Northern Territory case. Retrieved from http://beyondweird.com/ufos/Keith_Basterfield.Australian_Abduction_cases_2.html.

*Chalker, B. (1999). An interesting aside. Retrieved from http://www.cufos.org/IURspring 99 addendum.html

*Chalker, B. (1999). 1988, DNA sample from Khoury abduction raises big questions. Retrieved from http://www.ufocasebook.com/khouryabduction.html

*Chalker, B. & Harrison, D. (2001). The Gundiah, Mackay alien abduction: A preliminary report. Retrieved from http://www.ufocasebook.com/gundiahmackay.html

Childress, D. H. (2000). Technology of the gods: The incredible sciences of the ancients. Kempton, IL: Adventures Unlimited Press.

*Chudzinski, W. (2003). A collection of reports about paranormal events. In Piotr Cielebias' UFOs over Poland: The land of high strangeness (Chapter 18, pp. 129-131). West Yorkshire, England: Flying Disk Press.

*Clark, J. (1998). The UFO book: Encyclopedia of the extraterrestrial [Shane Kurz case] (pp. 536-537) Detroit, MI: Visible Ink Press.

Clark, J. (1998). The UFO encyclopedia: The phenomenon from the beginning (Vol. 1: A-K, p. 497), Detroit, MI: Omnigraphics.

*Clark, J. (1998). The UFO encyclopedia: The phenomenon from the beginning [Andreasson abduction case] (Vol. 1: A-K, pp. 86-95), Detroit, MI: Omnigraphics.

*Clark, J. (1998). The UFO encyclopedia: The phenomenon from the beginning [Antonio-Villas Boas case] (Vol. 2: L-Z, pp. 974-977), Detroit, MI: Omnigraphics.

*Clark, J. (1998). The UFO encyclopedia: The phenomenon from the beginning [Pascagoula abduction/Charles Hickson case] (Vol. 2: L-Z, pp. 714-719), Detroit, MI: Omnigraphics.

*Clark, J. (1998). The UFO encyclopedia: The phenomenon from the beginning [Carl Higdon case] (Vol. 2: L-Z, pp. 487-489), Detroit, MI: Omnigraphics.

*Clark, J. (1998). The UFO encyclopedia: The phenomenon from the beginning [Larson abduction case] (Vol. 2: L-Z, pp. 573-576), Detroit, MI: Omnigraphics.

*Clark, J. (1998). The UFO encyclopedia: The phenomenon from the beginning [Pascagoula abduction/Calvin Parker case] (Vol. 2: L-Z, pp. 714-719), Detroit, MI: Omnigraphics.

*Clark, J. (1998). The UFO encyclopedia: The phenomenon from the beginning [Duas Pontes abduction case] (Vol. 2: L-Z, pp. 342-343), Detroit, MI: Omnigraphics.

*Clark, J. (1998). The UFO encyclopedia: The phenomenon from the beginning [Patty Roach abduction case] (Vol. 2: L-Z, pp. 800-802), Detroit, MI: Omnigraphics.

*Clark, J. (1998). The UFO encyclopedia: The phenomenon from the beginning [David Seewaldt abduction case] (Vol. 1: A-K, p. 227), Detroit, MI: Omnigraphics.

*Clark, J. (1998). The UFO encyclopedia: The phenomenon from the beginning [Schirmer abduction case] (Vol. 2: L-Z, pp. 817-821), Detroit, MI: Omnigraphics.

*Clark, J. (1998). The UFO encyclopedia: The phenomenon from the beginning [Jose Antonio da Silva abduction case] (Vol. 1: A-K, pp. 149-152), Detroit, MI: Omnigraphics.

*Clark, J. (1998). The UFO encyclopedia: The phenomenon from the beginning [David Stephens abduction case] (Vol. 2: L-Z, pp. 685-690), Detroit, MI: Omnigraphics.

*Clark, J. (1998). The UFO encyclopedia: The phenomenon from the beginning [Whitley Strieber abduction case] (Vol. 2: L-Z, pp. 886-887), Detroit, MI: Omnigraphics.

*Clark, J. (1998). The UFO encyclopedia: The phenomenon from the beginning [Fred Valentich abduction case] (Vol. 2: L-Z, pp. 964-968), Detroit, MI: Omnigraphics.

*Clark, J. (1998). The UFO encyclopedia: The phenomenon from the beginning [Travis Walton abduction case] (Vol. 2: L-Z, pp. 981-998), Detroit, MI: Omnigraphics.

*CNNiReport (2012). Alien abduction experience in India. Retrieved from http://ireport.cnn.com/docs/DOC-838658

*Cole, W. (1977). Saucer-type 'ship' with multi-colored lights, landed in yard [William Cole case]. Retrieved from http://www.ufoevidence.org/cases/case332.html

*Collins, A. (1978). The Avely abduction case. Retrieved from http://www.ignaciodarnaude.com/ufologia/collins%20Abele y%20Abduction%201977,FSR78V23N6.pdf.

*Corrales, S. (2002). Electrician alleges UFO abduction in Southern Chile [Gabriel Encina case]. Retrieved from http://www.ufoevidence.org/cases/case58.html

*Corrales, S. (2004). The Acevedo abduction. Retrieved from http://www.ufoinfo.com/news/acevedo.shtml

*Cousineau, P. (1995). UFO secrets revealed (Chapter 5, Secrets in the saucers, pp. 101-102), New York, NY: Harper Collins West.

Craddock, P. Craddock, T., Gardner, J., Gardneer, M. (Executive Producers), Fox, J. (Director, Writer, and Narrator), Gardner, J. (Writer), Christopher, T. (Writer), FCZ media, LLC (Producer for History). (2009). I know what I saw [DVD]. Available from History Channel.

*Di Stefano, R. (2014). Chronicle of an incredible true story [The Zanfretta case]. Retrieved from http://www.rinodistefano.com

Dolan, R. (2002). UFOs and the national security state. Charlottesville, VA: Hampton Roads.

Donderi, D. (2013). UFOs, ETs, and alien abductions: A scientist looks at the evidence. Charlottesville, VA: Hampton, Roads Publishing.

*Domanski, G. (2000-2001). The multiple abductions of Mr. Wladyslav. Retrieved from www.thenightsky.org/wladyslav2000-2001.html

Druffel, A. (2003). Firestorm: Dr. James E. McDonald's Fight for UFO science. Columbus, NC: Wildflower Press.

*Feals, J. (2011). UFO fest attendees: 'We are not alone'[Hewins' twins' case]. Retrieved from www. Seacoastonline.com/article/20110904/News/109040337the Portsmouth Herald Portsmouth .

*Felber, R. (2015). Mohave incident (Chapters 1& 2: Tom & (Elise Gilford) case, p. 26; p. 31; pp. 39-44). Fort Lee, NJ: Barricade Books.

*Felber, R. (2015). Mohave incident (Chapters 1& 2: (Tom) & Elise Gilford case, p. 26; p. 31; pp. 39-44). Fort Lee, NJ: Barricade Books.

*Flying Saucer Review Publications, Ltd. (1981). Weird night and missing time on the Pennine Hills, United Kingdom [John & Michelle case]. Retrieved from http:// www.ufoevidence.org/cases/case29.html

Fowler, R. (2005). Allagash abductions: Undeniable existence of alien intervention [Chapter 3, Jim Weiner hypnosis sessions #1, pp. 35-52] (3rd. ed.). Columbus, NC: Wildflower Press.

*Fowler, R. (2005). Allagash abductions: Undeniable existence of alien intervention [Chapter 4, Jack Weiner case, pp. 67-101] (3rd ed.). Columbus, NC: Wildflower Press.

*Fowler, R. (2005). Allagash abductions: Undeniable existence of alien intervention [Chapter 5, Charles Foltz case, pp. 99-125] (3rd ed.). Columbus, NC: Wildflower Press.

*Fowler, R. (2005). Allagash abductions: Undeniable existence of alien intervention [Chapter 6, Chuck Rak case, pp. 126-150] (3rd ed.). Columbus, NC: Wildflower Press.

*Fowler, R. (2005). Allagash abductions: Undeniable existence of alien intervention [Chapter 7, Jack Weiner case, Not again!, pp. 151-180] (3rd ed.). Columbus, NC: Wildflower Press.

*Fuller, J. G. (1966). The interrupted journey: Two hours aboard a flying saucer [Barney Hill sessions 1, 2, & 3, pp. 69-131; 184-210]. London, England: Souvenir.

*Fuller, J. G. (1966). The interrupted journey: Two lost hours aboard a flying saucer [Betty Hill sessions 1, 2, & 3, pp. 135-223]. London, England: Souvenir.

*Garoutte, A. (2005). The Paris Colorado abduction case. Retrieved from Ufoexperiences.blogspot.com/2005/05/paris-colorado-abduction-case.html

*Georgian Ufological Association (1989). Contact case in Georgia (Eastern Europe) [David D. case]. Retrieved from http://www.ufoevidence.org/cases/case998.html

*AJ Gevaerd. (2006, December 28). UFO at Maringa.: UFO seduction case [man, 21] in Brazil. Retrieved from http://www.ufoexperiences.blogspot.com/2006/12/ufo-at-maringa.html

*AJ Gevaerd. (2006, December 28). UFO at Maringa.: UFO seduction case [teen, 13] in Brazil. Retrieved from http://www.ufoexperiences.blogspot.com/2006/12/ufo-at-maringa.html

*Granchi, I. (1977). Abduction by robot-like beings in Brazil [La Rubia case]. Retrieved from http://www.ufoevidence.org/cases/case464.html

*Green, G. (1965). Apolinar A. Villa case. Retrieved from http://www.ufoevidence.org/cases/case985.html

Green, S. B., Salkind, N. J., & Akey, T. M. (2000). Using SPSS for windows: Analyzing and understanding data. Upper Saddle River, NJ: Prentice Hall.

*Haines, R. F. (1999). CE5: Close encounters of the 5th kind [Ronneburg, Saxony Germany case, pp. 122-123]. Naperville, IL: Sourcebooks, Inc.

Hamilton, W. F., III (1996). Alien magic: UFO crashes, abductions, underground bases. Newbrunswick, NJ: Global Communications.

Hanshaw, L. G. (2004). Skepticism about selected paranormal events and why some believe we are not alone. Popular Culture Review, 15(2), 103-113.

Hill, P. R. (1995). Unconventional flying objects: A scientific analysis. Charlottesville, VA: Hampton Roads.

*Holly, C. (2008). Unexplained events of abduction and lost time-Long Island, New York. Retrieved from http://www.ufocasebook.com/2008c/losttimenewyork.html

Hopkins, B. (1981). Missing time. New York, NY: Ballantine Books.

*Hopkins, B. (1981). Missing time [Virginia Horton case, pp. 119-144]. New York, NY: Ballantine Books.

Hopkins, B. (1981). Missing time [Judy Kendall case, pp. 63-65]. New York, NY: Ballantine Books.

*Hopkins, B. (1981). Missing time [Steven Kilburn case, pp. 39-63]. New York, NY: Ballantine Books.

*Hopkins, B. (1981). Missing time [Dennis "Mac" McMahon case, pp. 101-105]. New York, NY: Ballantine Books.

*Hopkins, B. (1981). Missing time [David Oldham case, pp. 97-100]. New York, NY: Ballantine Books.

*Hopkins, B. (1981). Missing time [Phillip Osborne case, pp. 147-175]. New York, NY: Ballantine Books.

*Hopkins, B. (1981). Missing time [Howard Rich case, pp. 79-97]. New York, NY: Ballantine Books.

*Hopkins, B. (1987). Intruders: The incredible visitations at copley woods [Kathie Davis case, pp. 28-100]. New York, NY: Random House.

*Hopkins, B. (1992). The Linda Cortile abduction case. Retrieved from http://image. slidesharecdn.com/mufonufojournal-19929-september-130608181034-phpapp)...

*Hopkins, B, & Rainey, C. (2003). Sight unseen: Science, UFO invisibility and transgenic beings ["Karen" case, pp. 18-19]. New York, NY: Atria Books.

*Hopkins, B, & Rainey, C. (2003). Sight unseen: Science, UFO invisibility and transgenic beings [Lisa case, pp. 207-213]. New York, NY: Atria Books.

*Hopkins, B, & Rainey, C. (2003). Sight unseen: Science, UFO invisibility and transgenic beings [Marianne case, pp. 31-38]. New York, NY: Atria Books.

*Hopkins, B, & Rainey, C. (2003). Sight unseen: Science, UFO invisibility and transgenic beings [Melissa (& Will) case, pp. 340-356]. New York, NY: Atria Books.

*Hopkins, B, & Rainey, C. (2003). Sight unseen: Science, UFO invisibility and transgenic beings [Molly (& Danny) case, pp. 23-31]. New York, NY: Atria Books.

*Hopkins, B, & Rainey, C. (2003). Sight unseen: Science, UFO invisibility and transgenic beings [Danny (& Molly) case, pp. 23-31]. New York, NY: Atria Books.

*Hopkins, B, & Rainey, C. (2003). Sight unseen: Science, UFO invisibility and transgenic beings [Edward Reynolds case, pp. 258-259; 269-270; 273-275]. New York, NY: Atria Books.

*Hopkins, B. & Rainey, C. (2003). Sight unseen: Science, UFO invisibility and transgenic beings [Sally case, pp. 214-217; 228; 253-57]. New York, NY: Atria Books.

*Hopkins, B. & Rainey, C. (2003). Sight unseen: Science, UFO invisibility and transgenic beings [Will (& Melissa) case], pp. 340-356]. New York, NY: Atria Books.

*Hopkins, B. & Rainey, C. (2003). Sight unseen: Science, UFO invisibility and transgenic beings [Katharina (Wilson) case], pp. 340-356]. New York, NY: Atria Books.

*Hopkins, B. & Rainey, C. (2003). Sight unseen: Science, UFO invisibility and transgenic beings [Terry Winthrop case], pp. 188-206]. New York, NY: Atria Books.

Hopkins, B. (2005). The faith-based science of Susan Clancy. Retrieved from http://www.unknowncountry.com/insight/faith-based-science-susan-clancy-budd-hopkins.

*Howe, L. M. (1994). Glimpses of other realities: Vol. I: Facts & eyewitnesses [Judy Doraty case, pp. 198-202]. Cheyenne, WY: Pioneer Printing.

*Howe, L. M. (1994). Glimpses of other realities: Vol. I: Facts & eyewitnesses [Cindy Tindle case, pp. 202-226]. Cheyenne, WY: Pioneer Printing.

*Huyghe, P. (1996). The field guide to extraterrestrials: A complete overview of alien life forms-based on actual accounts and sightings [Harrison Bailey case, pp. 86-87]. New York, NY: Avon Books.

*Huyghe, P. (1996). The field guide to extraterrestrials: A complete overview of alien life forms-based on actual accounts and sightings [male soldier's case, pp. 34-35]. New York, NY: Avon Books.

*Huyghe, P. (1996). The field guide to extraterrestrials: A complete overview of alien life forms-based on actual accounts and sightings [Mike Shea case, pp. 90-91]. New York, NY: Avon Books.

*Huyghe, P. (1996). The field guide to extraterrestrials: A complete overview of alien life forms-based on actual accounts and sightings [Fred Reagan case, pp. 110-111]. New York, NY: Avon Books.

*Huyghe, P. (1996). The field guide to extraterrestrials: A complete overview of alien life forms-based on actual accounts and sightings [Jamie W. case, pp. 18-19]. New York, NY: Avon Books.

Hynek, J. A. (1972). The UFO experience: A scientific inquiry. New York, NY: Ballantine Books.

Hynek, J. A. (1997). The Hynek UFO report. New York, NY: Barnes & Noble.

*Imbrogno, P. J. (1987). Contact of the fourth kind (abductions in the Hudson Valley area of New York) [Bill's case]. Retrieved from http://www.skepticfiles.org/ufo1/hudson.html

*Imbrogno, P. J. (1987). Contact of the fourth kind (abductions in the Hudson Valley area of New York) [Gail case]. Retrieved from www.unexplainable.net/abductions_in_the_hudson_valley_area_2639.php

*Inexplicata. (2011). Abduction experiences in Latin America [Rolando Quiroga Valero case]. Retrieved from http://www.ufoevidence.org/documents/doc1859.html

*Internet Broadcasting Systems and Local6.com. (2005). Couple claims aliens abducted, probed them [Clayton Lee case]. Retrieved from http://www. ufocasebook.com/leeabduction.html.

Jacobs, D. M. (1992). Secret life: Firsthand documented accounts of UFO abductions. New York, NY: Simon & Schuster.

*Jacobs, D. M. (Ed.). (2000). UFOs & abductions: Challenging the borders of knowledge [Karen (& Richard) case, pp. 230-231]. Lawrence, Kansas: University Press of Kansas.

*Jacobs, D. M. (Ed.). (2000). UFOs & abductions: Challenging the borders of knowledge [Richard (& Karen) case, pp. 230-231]. Lawrence, Kansas: University Press of Kansas.

Jacobs, D. M. (2006). A review of abducted: How people Come to believe they were kidnapped by aliens by Susan Clancy. Retrieved 7/10/2018 from http://www. ufoabduction.com/clancy.review.html

*Jasek, M. Abduction on the North Canol Road, Canada [Kevin's account]. Retrieved from http://www.ufobc.ca/yukon/n-canol-abd/index.html

Jet Propulsion Laboratory (November 7, 2003). Voyager: Celebrating 25 years of discovery. Retrieved November 11, 2003,from http://www.voyager.jpl.nasa.gov/

Jet Propulsion Laboratory (2015). Voyager: The golden record. Retrieved January 15, 2017 from http://www.voyager.jpl.nasa.gov/spacecraft/goldenrec.html

Kean, L. (2010). UFOs, generals, pilots, and government officials go on the record. New York, NY: Three Rivers Press.

Kean, L. 92011). Special presentation secret access: UFOs on the record (Stern, R., and Sudberg, A., Producers, Filmmakers). [DVD]. Available from the History Channel.

*Kessler, D. (2010). Thomas Reed family abduction. Retrieved from http://www. ufocasebook.com/2010/reedabduction.html

Koenker, R. H. (1974). Simplified statistics. Totowa, NJ: Littlefield, Adams & Co.

*Kolden, J. (1992). UFO abducts 4 children. Retrieved from http://www.greatdreams. com/ufos/kids_abducted.html

*Kwek, G. (2010). Out of this world: Russian region leader's alien abduction story shakes officials [Kirsan Ilyumzhinov case]. Retrieved from http;//www.thefw.com/ famous-alien-abduction.

Laibow, R. L. (1993). Clinical discrepancies between expected and observed data in patients reporting UFO abductions: Implications for treatment, pp. 1-10. Retrieved April 3, 2016, from http://www.ufoevidence.org/documents/doc7.html

Linn, R. L. & Gronlund, N. E. (1995). Measurement and assessment in teaching. (7th ed.). Upper Saddle River, NJ: Prentice Hall.*

*Lorenzen, C. & Lorenzen, J. (1976). Encounters with UFO occupants [Gerry Irwin case, pp. 347-356]. New York, NY: Berkley Publishing Corporation.

*Lorenzen, C. & Lorenzen, J. (1976). Encounters with UFO occupants [Onilson Patero case, pp. 356-361]. New York, NY: Berkley Publishing Corporation

*Lorenzen, C. & Lorenzen, J. (1976). Encounters with UFO occupants [Paolo Caetano Silveira case, pp. 165-166]. New York, NY: Berkley Publishing Corporation.

Lynn, S. J. & Kirsch, I. I. (1996). Alleged alien abductions: False memories, hypnosis, and fantasy proneness. Psychological Inquiry 7(2), 151-155.

Maccabee, B. (1987). The fantastic flight of JAL1628. Retrieved July 16, 2016 from http://brumac.8k.com/JAL1628/JL1628.html

Maccabee, B. (2014). The FBI-CIA-UFO connection. Charlottesville, VA: Richard Dolan Press.

*Mace, C. (2007). Abduction of Helene Guilana by aliens, June 11, 1976. Retrieved from http://www.ufodigest.com/news/0807/helene.html

*MacGregor, T. & MacGregor, H. (2013). Aliens in the backyard: UFO encounters, abductions & synchronicity [Connie J. Cannon case, pp. 34-44]. Hertford, NC: Crossroads Press.

*MacGregor, T. & MacGregor, H. (2013). Aliens in the backyard: UFO encounters, abductions & synchronicity [Diane Fine case, pp. 78-81]. Hertford, NC: Crossroads Press.

Mack, J. E. (1994). Abduction: Human encounters with aliens. New York, NY: Scribner's Sons.

*Mack, J. E. (1994). Abduction: Human encounters with aliens [Authur's case, pp. 369-386]. New York, NY: Scribner's Sons.

*Mack, J. E. (1994). Abduction: Human encounters with aliens [Carlos' case, pp. 335-368]. New York, NY: Scribner's Sons.

*Mack, J. E. (1994). Abduction: Human encounters with aliens [Catherine's case, pp. 143-176]. New York, NY: Scribner's Sons.

*Mack, J. E. (1994). Abduction: Human encounters with aliens [Dave's case, pp. 265-292]. New York, NY: Scribner's Sons.

*Mack, J. E. (1994). Abduction: Human encounters with aliens Ed's case, pp. 51-68]. New York, NY: Scribner's Sons.

*Mack, J. E. (1994). Abduction: Human encounters with aliens [Eva's mission, pp.241-263]. New York, NY: Scribner's Sons.

*Mack, J. E. (1994). Abduction: Human encounters with aliens [Jerry's case, pp. 117-118]. New York, NY: Scribner's Sons.

*Mack, J. E. (1994). Abduction: Human encounters with aliens [Joe's case, pp. 177-200]. New York, NY: Scribner's Sons.

*Mack, J. E. (1994). Abduction: Human encounters with aliens [Paul's case, pp. 217-240]. New York, NY: Scribner's Sons.

*Mack, J. E. (1994). Abduction: Human encounters with aliens [Peter's case, pp. 293-334]. New York, NY:Scribner's Sons.

*Mack, J. E. (1994). Abduction: Human encounters with aliens [Sara's case, pp. 201-216]. New York, NY: Scribner's Sons.

*Mack, J. E. (1994). Abduction: Human encounters with aliens [Scott's case, pp. 91-109]. New York, NY: Scribner's Sons.

*Mack, J. E. (1994). Abduction: Human encounters with aliens [Sheila N. case, pp. 69-90]. New York, NY: Scribner's Sons.

*Marden, K. & Stoner, D. (1988). Alien abduction files [Jennie Henderson case, pp. 131-214]. Pompton Plains, NJ: New Page Books.

*Marden, K. & Stoner, D. (1988). Alien abduction files [Andrew Stevens case, pp. 214-216]. Pompton Plains, NJ: New Page Books.

*Marden, K. & Stoner, D. (1988). Alien abduction files [Denise Stoner case, pp. 45-70]. Pompton Plains, NJ: New Page Books.

*Martinez, J. R. (1975). Another Bahia Blanca case [Carlos Alberto Diaz case]. Retrieved from http://www.ufoevidence.org/cases/case307.html

*Maurer, D. A. (1990). Doctor says aliens took her on UFO [Rauni-leena Luukanen case, pp. 1-3]. Retrieved from http://www.textfiles.com/ufo/SIGHTINGS/ufoncs19.txt

*McCombs, P. (1987). Ufo 'Abductees' dare to bare all [Shirley Coyne case]. Orlando Sentinel. Retrieved from http://www.articles.orlandosentinel.com15/lifestyle/0180250 056_1_ab -abduction

Mill Creek Entertainment (2006). Aliens, abductions, & extraordinary sightings: A paranormal documentary collection [Travis Walton segment]. Available from Millcreekent.com

*Mitchell, B. (2015). What if [Josie case, pp. 227-246]? Lexington, KY: Author

*Mitchell, B. (2015). Intrusion: Alien encounters [Maryanne case, pp. 209-212]. Lexington, KY: Author.

*Mitchell, B. (2015). Intrusion: Alien encounters [Kellie Mclean case, pp. 50-62]. Lexington, KY: Author.

*Mizera, M. (1981). Lintel Lake. Germany incident. Retrieved from http://www.ufocasebook.com

*Morphy, R. (2011). Giant space brains of Palos Verdes (CA) [Peter Rodriguez & John Hodges case, pp. 1-6]. Retrieved from http://www.cryptopia.us/site/2011/01/giant-space-brains-of-palos-verdes-california-usa/

*Murphy, B. (1988). Alien abductions [Marianne Shenefield case]. Retrieved from http://www.textfiles.com/ufo/SIGHTINGS/abductee.ufo

*Nagaitis, C., & Mantle, P. (1953). 1953: July UFO & alien sightings [Gerry Armstrong case, pp. 1-2]. Retrieved from http://www.thinkaboutitdocs. com/1953-july-ufo-alien-sightings/

*National UFO Reporting Center (2004). Close-range sighting of circular craft; missing time. Retrieved from http://www.ufoevidence.org/cases/case1064.html

Pallant, J. (2010). Survival manual: A step by step guide to data analysis using SPSS. (4th ed.). NewYork, NY: McGraw-Hill.

*Pomorski, H. & Adamczyk, K. (1978). Abductee Jan Wolski interviewed: 30th anniversary of Emilcin abduction. Retrieved from http://www.ufocasebook. com/2008/wolski.html

*Porter, L. (1991). Linda Porter case. Retrieved from http://www.disclose.tv/forum/ highest-strangeness-body-containers-t16644.html

Puthoff, H. E. (1998). Can the vacuum be engineered for spacecraft applications?: Overview of theory and experiments. Journal of Scientific Exploration, 12(1), 295-308.

*Randles, J. (1980). Police constable Alan Godfrey's abduction in West Yorkshire, England. Retrieved from http://www.ufoevidence.org/cases/case722.html

*Randles, J. (1994). Alien contacts & abductions: The real story from the other side [Alfred Burtoo case, pp. 157-158]. New York, NY: Sterling Publishing Co., Inc.

*Raynes, B. (1965). Sid Padrick case. Retrieved from http://ufoevidence.org/cases/ case963.html.

*Robertson, D. (2000). 1978 Tim Cullen abduction. Retrieved from http://www. ufocasebook.com/1978cullen.html

*Malcolm Robinson. (2009, December 19). UK-UFO: A70 abduction case (Scotland) [Colin Wright regression session]. Retrieved from http://blog.ufo-blog. com/2009/12/uk-ufo-a70-abduction-case-scotland.html

*Romanek, L. (2014, August 14). Brian Scott: a remarkable abduction case. Retrieved from lisaromanekufoinsider.blogspot.com/2014/08brian-scott-remarkable-abductee-case.html

*Rosales, A. (1996). 1971-Abductee involved in UFO crash near Edwards AFB [Lori Cordini case]. Retrieved from www.ufocasebook.com

*Rosales, A. (1996). Louis Boisvert case. Retrieved from http://www.iraap.org/ rosales/1996.html

*Rosales, A. (1996). Norberto Perez case. Retrieved from http://www.iraap.org/ rosales/1996.html

*Rosales, A. (1996). Woman, North Yorkshire, England case.Retrieved from http:// www.iraap.org/rosales/1996.html

*Rosales, A. (2016). Humanoid encounters: The others among us [male Moroccon, 23 case, pp. 67-69]. Lexington, KY: Triangulum Publishing.

RTL Television, Belgium. (July 26, 2011). Wait a minute: Was the Belgian UFO picture really debunked? Retrieved from www.abovetopsecret.com/forum/ thread305202/pg1.

*Rueckert, C. & Elkins, D. (1977). Abduction by machine-like beings [Lee Parish case]. Retrieved from http://www.ufoevidence.org/cases/case354.html

Salkind, N. J. (2008). Statistics for people who (think they) hate statistics. (3rd ed.). Los Angeles, CA: sage Publications.

Salvia, J., Ysseldyke, J. E., & Bolt, S. (2007). Assessment in special and inclusive education, 10th ed., Boston, MA: Houghton Mifflin Company.

*Sanchez-Ocejo, V. (January 3,1979). Full synopsis of Filiberto Cardenas case, pp. 1-4. Retrieved from http://www.ufoeveidence.org/Cases/CaseSubarticle.asp?ID= 696

*Sanchez-Ocejo, V. (February 21,1979). Filiberto Cardenas case: Second encounter, pp. 3-4. Retrieved from http://www.ufoeveidence.org/Cases/CaseSubarticle. asp?ID= 696.http://www.ufoevidence.org/cases/case695.html

Sanderson, I. T. (1970). Invisible residents. Cleveland, OH: World Publishing Co.

*Sierra, R. (2008). UFO and alien encounters [Reuben Sierra case]. Retrieved from http://www.ufocasebook.com/2008c/ufoalienencounters.html

*Simondini, A. P., & Lecomte, D. (2014, January 28). Argentina: Police report documents 1978 UFO teleportation [Carlos Acevedo & Hugo Prambs case]. Retrieved from inexplicata.blogspot.com/2014/01/argentina-police-report-documents-1978.html

*Spencer, J. (1992). World atlas of UFOs: Sightings, abductions and close encounters [The Anders encounter case, pp. 103-104]. New York, NY: Smithmark Publishers.

*Spencer, J. (1992). World atlas of UFOs: Sightings, abductions and close encounters [Medinaceli abduction: The Julio Fernandez case, p. 108]. New York, NY: Smithmark Publishers.

*Spencer, J. (1992). World atlas of UFOs: Sightings, abductions and close encounters [The Beit Bridge encounter: Peter & Frances MacNorman case, pp. 149-151]. New York, NY: Smithmark Publishers.

*Spencer, J. (1992). World atlas of UFOs: Sightings, abductions and close encounters [The Maureen Puddy case, pp. 166-168]. New York, NY: Smithmark Publishers.

*Spencer, J. (1992). World atlas of UFOs: Sightings, abductions and close encounters [The Elias Seixas case, p. 188]. New York, NY: Smithmark Publishers.

*Stokes, R. (2012). Abduction in the desert [Antique dealer case]. Retrieved from http://www.ufocasebook.com/2012/abductiondesert.html

Stonehill, P., Mantle, P. (2016). Russia's USO secrets: Unidentified submersible objects in Russian and international waters. Charlottesville, VA: Richard Dolan Press.

*Story, R. D. (2001). The mammoth encyclopedia of extraterrestrial encounters [Tom Dawson encounter, pp. 180-181]. London, England: Constable & Robinson.

*St. Patrick's chair and well [Lawrence John case]. United Kingdom UFO Bulletin (1997). Retrieved from http://www.mysteriousbritain.co.uk/northern-ireland/ancient-sites/st-patrick%E2%80%99s... Sturrock, P. A. (1999). The UFO enigma: A review of the physical evidence. New York, NY: Warner Books.

*Takanashi, J.-I. (1979). Amano abduction. Retrieved from http://ufologie.patrickgross.org/ce3/1978-10-03-japan-sayamacity.html

*The Tujunga Canyon abductions (1953). [Sara Shaw & Jan Whitley case] . Retrieved from http://the nightsky.org/tujunga.html

*Think About It (1996, December, 9). [Plinio Bragatto abduction case]. Retrieved from http://www.thinkaboutitdocs.com/1996-november-december-ufo-alien-sightings/

Thorndike, R. M. & Dinnel, D. L. (2001). Basic statistics for the behavioral sciences. Upper Saddle River, NJ: Merrill Prentice Hall.

*Tier, Germany abduction (1978, November 25). [Pregnant woman case]. Retrieved from http://www. the cid.com/ufo/uf15/ufo/150764.html

*Tosti, J. (2005). The John Tosti story. Retrieved from http://www.ufoinfo.com/news/johntosti2.shtml

*Turner, K. (2013). Taken: Inside the alien-human abduction agenda [Angie's case, pp. 146-170]. Lexington, KY: Word Mean Publishing.

*Turner, K. (2013). Taken: Inside the alien-human abduction agenda [Anita's case, pp. 81-95]. Lexington, KY: Word Mean Publishing.

*Turner, K. (2013). Taken: Inside the alien-human abduction agenda [Beth's case, pp. 107-115]. Lexington, KY: Word Mean Publishing.

*Turner, K. (2013). Taken: Inside the alien-human abduction agenda [Jane's case, pp. 122-145]. Lexington, KY: Word Mean Publishing.

*Turner, K. (2013). Taken: Inside the alien-human abduction agenda [Lisa's case, pp. 66-80]. Lexington, KY: Word Mean Publishing.

*Turner, K. (2013). Taken: Inside the alien-human abduction agenda [Pat's case, pp. 26-30]. Lexington, KY: Word Mean Publishing.

*UFOinfo.com (1996). Israeli man shanghaied into space. Retrieved from http://www.ufoinfo.com/roundup/v01/rnd01 30.shtml

*UFOinfo.com (1970). Raymond Shearer case. Retrieved from http://www.ufoinfo.com/humanoid/humanoid1970.shtml

*UFO Related Entities Catalog (URECAT) (1978). Francisco Nunez case. Retrieved from http://www.ufologie.patrickgross.org/ce3/1978-07-06-argentina-mendoza.html

*Vike, B. (1990). 1990 Westchester, NY abduction. Retrieved from http://www.ufocasefiles.com

*Vike, B. (2007). Missing time on national airline from Las Vegas, Nevada to DFW. Retrieved from http://www.ufoinfo.com/news/missingtime.shtml

Vogt, P. W. (1999). Dictionary of statistics & methodology: A non-technical guide for the social sciences (2nd ed.). Thousand Oaks, CA: Sage Publications.

Welles, H. (1981). Monroe, NC man is abducted and has meeting with occupant of Craft [Pat Eudy case]. Retrieved from http://www.ufoevidence.org/cases/case938.html

*Whiting, F. (1980). The abduction of Harry Joe Turner. Retrieved from http://www.slideshare.net/mufonnexus/mufon-ufo-journal-1980-3-march

Index

Index contains only references cited in text. Citations in References show cases and their use in this study.

About the Author

Larry G. Hanshaw

Professor Emeritus of Secondary Science Education

Oxford, MS 38655

lhanshaw@bellsouth.net

662-234-2652 (H)

EDUCATION AND PROFESSIONAL CREDENTIALS

DEGREES

B. S., 1969	Chemistry	Tougaloo College Tougaloo, MS
M. Ed., 1974	Natural Science (Chemistry) & Education	University of Delaware Newark, Delaware
Ph.D., 1976	Science Education (Chemistry)	University of Southern MS Hattiesburg, MS

OTHER TRAINING

System Energy Resources, Inc., Grand Gulf Nuclear Station Port Gibson, MS, "Boiling Water Reactor Radio Chemistry and Plant Systems Course", 1/87-5/87; Certificate; Radio Chemist II

PROFESSIONAL EXPERIENCE

2007 - 2014   Professor of Secondary Science Education
School of Education, University of Mississippi
University, Mississippi

1989 - 2007   Associate Professor of Secondary Science Education
School of Education, University of Mississippi
University, Mississippi

1988 - 1989   Associate Professor of General Science
Jackson State University, Jackson, Mississippi

1986 - 1988   Radio Chemist II, System Energy Resources, Inc.
Grand Gulf Nuclear Station, Port Gibson, Mississippi

1985 - 1986   Associate Professor of Physical Science/Chemistry
Alcorn State University, Lorman, Mississippi

1976- 1985   Assistant Professor of Physical Science/Chemistry
Alcorn State University, Lorman, Mississippi

1974 - 1976   Teaching Assistant
Department of Science Education
University of Southern Mississippi
Hattiesburg, Mississippi

1972 - 1974   General Science Teacher and Planetarium Operator
Bayard Middle School
Wilmington, Delaware

1969 - 1970   Chemist/Technical Salesman
E. I. Dupont deNemours and Company, Inc.,
Chestnut Run Laboratories
Wilmington, Delaware

Professional Honors

2014   **Awarded Honor of Professor Emeritus of Secondary Science Education**, School of Education, University of Mississippi; **Distinguished Service Citation**, Chancellor Daniel W. Jones, May.

2007   **Promoted to rank of Professor**, School of Education, University of Mississippi, May.

1996	**Selected to intern as an Assistant Commissioner for Academic Affairs**, MS IHL, January-June.
1995	**Achieved tenure**; School of Education, the University of Mississippi, May.
1985	**Promoted to rank of Associate Professor**, Department of Chemistry and Physics, Alcorn State University, Lorman, Mississippi. August.
1985;1986	**Summer Faculty Research Appointments**, Environmental Research Division, Argonne National Laboratory, Argonne, IL. June-August; July-August.
1984	**Summer Faculty Research Appointment**, Applied Science Division, Brookhaven National Laboratory, Upton, L.I., NY. May-August.

Professional Publications

Doctoral Dissertation

Hanshaw, L. G. (1976). Test anxiety, self-concept, and the test performance of students paired for testing and the same students working alone. Dissertation Abstracts International, 37 (10), 6387-A. (University Microfilms No. 77, 5945).

Book

Hanshaw, Larry G. (February, 2006). Cooperative classroom testing. Baltimore, MD: University Press of America.

Articles in Refereed Journals

Hanshaw, L. G. (1982). Test anxiety, self-concept, and the test performance of students paired for testing and the same students working alone. Science Education, 66 (1),15-24.

Hanshaw, L. G. (1991). General procedures for identification and diagnosis of mildly handicapped and at-risk children. National Forum of Special Education Journal, 2(1), 30-38.

Hanshaw, L. G. (1993). New basic science programs raise requirements and enhance the preparation of secondary teachers. Journal of the Mississippi Academy of Sciences, 38(3), 5-7.

Hanshaw, L. G. (1994). Expanding the levels of teacher-selected questions on chemistry examinations. Journal of the Mississippi Academy of Sciences, 39(2), 3-7.

Hanshaw, L. G. (1994). Utilizing class assignments to enhance the curriculum perspectives of secondary teachers. National Forum of Special Education Journal, 41(1&2), 15-21.

Hanshaw, L. G. (1995). The Literary influence of alchemy: An explication of selected works. Popular Culture Review, 6(1), 61-73.

Hanshaw, L. G. (1996). The inaccessible die: An activity integrating inquiry teaching and problem solving. Journal of the Mississippi Academy of Sciences, 41(4), 158-165.

Love, F. E., Henderson, D., & Hanshaw, L. G. (1996). Preparing pre-service teachers to understand diversity in classroom management. College Student Journal, 30(1),112-118.

Blackbourn, J. M. (Hanshaw, L. G., Hamby, D., & Beck, J. B. (1996-97). The total quality curriculum. National Forum of Applied Educational Research Journal, 10(1), 24-30.

Hanshaw, L. G. (1998). Analyzing the journal entries of student teachers. The Professional Educator, 20(2), Spring, 31-44.

Hanshaw, L. G. (2004). Value-related issues in a departmental merit pay plan. The Professional Educator, 26(2), Spring, 57-68.

Hanshaw, L. G. (2004). Skepticism about selected paranormal events and why some believe we are not alone. Popular Culture Review, 15(2), September, 103-113.

Hanshaw, L. G. (2007). Addressing the need for appropriate science courses in graduate secondary education programs [Abstract]. 71st Annual MS Academy of Science Journal, 52(1).

Hanshaw, L. G., Williams-Black, T., Boyd, N., Smothers-Jones, B., Love, F., Thompson, J. (2010). Examining the conflict resolution modes of clinical supervisors and teacher education candidates. College Student Journal, 44(2), 250-265.

Hanshaw, L. G. (2012). Qualitative aspects of group-only testing. College Student Journal, 46(2), June, 419-426.

Research Reports

Hanshaw, L. G. (1982). "Correlations Between Major Atmospheric Deposition Constituents and Selected Soil Parameters in Loblolly Pine Forests". Unpublished report.

Hanshaw, L. G. (1984). "A Determination of PPB Levels of Formaldehyde in Ambient Air and Rainwater". Collaborative Research Projects as a Visiting Assistant Chemist with Dr. Roger L. Tanner; Official Informal Report, Brookhaven National Laboratory, Upton, L. I., NY, 22 pp.

Hanshaw, L. G. (1986). "A Field Evaluation of the Affects of Ozone and Moisture Stress on Soybeans and Sulfur Dioxide and Ozone Interaction on Corn." Research Appointment Summary Report with Dr. Patricia M. Irving, Senior Research Ecologist, Argonne National Laboratory, Argonne, IL, June-August, 1985; August, 1986, 12 pp.

Grant Awards

1986   **Department of Energy Used Instrumentation/ Equipment Grant, Principal Writer**, Department of Chemistry, Alcorn State University, Lorman, MS. August, $10,000.

1982-1985 **Computer-Assisted Instructional Program for Non-Science Majors**, Co-Director, Title III Grant, Alcorn State University, Lorman, MS. October, $120,000.

1991   **Departmental Award, Secondary Science Resources Center**, Department of Curriculum and Instruction. Purchased equipment, supplies, etc., and loaned same to science teachers to facilitate science instruction at area schools. University of Mississippi, $3,000.

1991-1992 **Mississippi Alliance for Minority Participation**, SiteCoordinator (University of Mississippi); a $10-million National Science Foundation grant to all eight IHL institutions to increase minority participation in mathematics, science, science education and engineering; attended meetings and contributed to writing of grant. Over $300,000 to the University of MS's MAMP Programs.

1992   **Partners and Associates Grant**, Computer equipment for educational research, University of Mississippi, $3407.

Course Teaching Experiences

Curriculum	Education/Methods	Science
EDCI 351	**EDSE 400**	General .Chemistry
EDCI 503*	**EDSE 445**	Physical Science
EDCI 601*	**EDSE 645**	General Science

Statistics & Educational Research	Educational Psychology	Research Writing
EDRS 601	**EDPY 303***	**EDUC 555***
EDRS 605	**EDPY 307**	

Note: Courses in bold taught at University of MS; w/asterisk were courses I redesigned/developed.

www.ingramcontent.com/pod-product-compliance
Lightning Source LLC
Chambersburg PA
CBHW051411200326
41520CB00023B/7199